Purposely Profitable

Purposely Profitable

Embedding Sustainability into the DNA of Food Processing and Other Businesses

Brett Wills

WILEY Blackwell

Library of Congress Cataloging-in-Publication Data

Names: Wills, Brett, author.
Title: Purposely profitable : embedding sustainability into the DNA of food processing and other businesses / Brett Wills.
Description: Hoboken, NJ : John Wiley & Sons, 2016. | Includes bibliographical references and index.
Identifiers: LCCN 2015040311 (print) | LCCN 2015042193 (ebook) | ISBN 9781118978153 (pbk.) | ISBN 9781118978139 (pdf) | ISBN 9781118978146 (epub)
Subjects: LCSH: Organizational effectiveness. | Food industry and trade. | Sustainability. | Strategic planning.
Classification: LCC HD58.9 .W556 2016 (print) | LCC HD58.9 (ebook) | DDC 664.0068/4–dc23
LC record available at http://lccn.loc.gov/2015040311

A catalogue record for this book is available from the British Library.

*To my beautiful and loving wife Agnieszka, without you,
none of this would be possible.*

Contents

About the Author

Brett Wills has over 15 years of experience in the area of sustainable business management and coaches organizations across the globe in the development and execution of winning strategies that realize sustainable performance.

Brett established his career as the Plant Manager for a leading technology company based in Toronto, Canada. He went on to establish the Green Enterprise Movement Inc., a firm dedicated to helping organizations optimize performance through a focus on integrating sustainability into the business strategy. Brett also serves as the Director of Sustainability for High Performance Solutions Inc., an organization focused on driving business performance through collaboration. He is the lead facilitator of two sustainability collaboratives in North America's largest Eco-Business zone where he works with leading organizations to accelerate performance through a collaborative approach to sustainability.

As a Professor in the Green Business Management graduate program at Seneca College in Toronto, Canada, Brett is helping to prepare the next wave of Sustainability change agents to enter the workforce.

A sought after speaker in the area of sustainability and strategy, he is also the author of *Green Intentions: Creating a Green Value Stream to Compete and Win*, a book that presents a systematic process known as "Lean and Green" to support organizations in their efforts to systematically and profitably improve environmental performance. Brett lives in Toronto with his wife Agnieszka where he volunteers his time to mentor young entrepreneurs.

Connect with Brett on Twitter @greenintentions

Learn more at www.greenenterprise.ca

Sustainability Primer

Understanding sustainability

Viewed by many as one of the largest movements in history, Sustainability means many things to many people. Ask the mayor of a city and they may talk about reducing the social and environmental impacts of urban development, ask a farmer and they may talk about environmentally friendlier agricultural practices or the importance of local food, ask a corporate executive and they may talk about energy conservation or strategic philanthropy.

Because Sustainability means many things to many people, it can be challenging to understand. Why are there so many views or "definitions" of Sustainability? The answer is quite simple but requires an understanding of the basic, fundamental concept of Sustainability which is to balance and continually improve social, environmental and economic performance. Recognizing this basic concept leads to understanding how a farmer would define Sustainability differently than an executive or mayor since the focus for improving social, environmental and economic performance will vary based on the end goal. For example, a farmer may be focused on the long term health of soil, where a corporate executive may be focused on energy conservation. At the end of the day, while the detailed description of how to achieve Sustainability may differ depending on who is asked, the basic concept and end game remains the same.

Given the focus of this book is on enhancing organizational performance, the concept of Sustainability as presented in this Primer and throughout other chapters, refers to Sustainability as it applies to organizations, regardless of the type of organization. The organization may be a private business, non-profit organization, even government organization.

To clearly define Sustainability as it applies to an organization, it's easier to start with a broad definition and distill this down to provide a more specific understanding. At a holistic, global level, one of the most popular definitions of Sustainability comes from the Brundtland Report (WCED, 1987) and defines Sustainability as "The ability to meet the needs of today's generation without compromising the ability of future generations to meet their own needs". While this provides a good start to defining Sustainability at a macro level, it leaves many questions unanswered when it comes to Sustainability on a more micro of firm level.

Fortunately, in the mid-1980s another widely accepted definition of Sustainability emerged when an organization called The Natural Step (TNS) was founded by a Swedish Doctor – Karl Henrik Robert. TNS set out to further clarify the meaning of Sustainability. In their efforts, TNS realized it was incredibly difficult to define what Sustainability is since it can mean so many things to so many people and thus, took a reverse angle to defining Sustainability. Instead of defining what Sustainability *is*, TNS asked what cannot happen if a system is to be sustainable? With this thinking, TNS developed four system conditions that must be satisfied in order for a system to be sustainable (www.thenaturalstep.org); let's explore these further.

The Natural Step four system conditions for sustainability
System condition no. 1
Resources from the Earth's crust (i.e. oil, copper, etc.) cannot be extracted faster than they can be regenerated. Since resources in the Earth's crust take thousands of years to develop, they are, for all intents and purposes, non-renewable. Like anything finite, continuing to consume these resources will eventually lead to their depletion which, in the long term, is unsustainable. This suggests that extraction of resources from the Earth's crust must stop (or dramatically slow down) and practices must migrate to more renewable and sustainable solutions. How to do this becomes a focus of Sustainability efforts based on the types of resources being consumed.

System condition no. 2
Toxins and pollutants cannot be discharged into the atmosphere faster than the earth can clean or otherwise deal with them. Picture a bubble that is closed off from everything around it (similar to Earth's closed atmosphere). Inside this bubble, there is one smoke stack and one tree. If the tree can clean 1 tonne of CO2e out of the air every year and the

smoke stack discharges 1 tonne of CO2e every year, there is a balance inside that bubble that can be sustained over time. However, once the stack discharges more than 1 tonne of CO2e, balance is disrupted, a build-up of toxins and pollutants occurs, resulting in environmental destruction, which is unsustainable over the long term. The point being, that while some "pollution" can be cleaned and tolerated by the Earth, it must be at a level that is sustainable. Additionally, there can be no toxins or pollutants discharged into the atmosphere or the environment that the Earth cannot deal with or handle, such as some fire retardants. Since these types of toxins and pollutants cannot be broken down by the environment, they stick around forever resulting in environmental destruction, which over the long term is also unsustainable. Again, how this is achieved becomes the focus of Sustainability efforts based on the types of toxins and pollutants be discharged.

System condition no. 3

Surface level resources (both on land and underwater) cannot be destroyed faster than they can be regenerated. Take a lake, for example, if fish in the lake are harvested faster than they regenerate, eventually there will be no fish left. Not only is this unsustainable for the specific species of fish, it can have a negative ripple effect on other species and the overall eco-system of which they belong. The point here is that the harvesting or destruction of surface level resources need not necessarily stop. Rather it needs to be done at a rate that allows for regeneration enabling continual use or harvesting over the long term, making it a sustainable source of resources. Once again, how this is achieved becomes the focus of Sustainability efforts based on the types of surface level resources being harvested or destroyed.

These first three system conditions are clearly focused on the environmental dimension of Sustainability, while the following fourth and final system condition looks at the higher level social and economic dimensions.

System condition no. 4

The world's resources cannot be inequitably distributed such that basic human needs are not met for all people around the world. In other words, social and economic systems must ensure at least the basic human needs of food, shelter, and water have been met for all people on Earth. If people do not have their basic human needs met, people focus solely on meeting those basic human needs and nothing else (i.e. reducing environmental impact). Therefore the social and economic conditions of the system must operate in a way that, at a minimum, covers the basic needs of people within that system. Now, this is where the definition of Sustainability differs between say a country or city vs that of a business or other type of organization.

This definition offered up by The Natural Step provides a much clearer picture of what Sustainability means at a broad, macro level. Clearly, satisfying these four system conditions would allow "current generations to meet their own needs without compromising the ability of future generations to meet their needs." The problem here is that this high-level picture of Sustainability is so broad that it becomes challenging to apply this at the macro of firm level, particularly when it comes the fourth system condition relating to the social and economic dimensions of Sustainability.

The sustainable organization

Building on the broader definitions provided by Brundtland and The Natural Step, organizations around the world began asking how the concept of Sustainability could be more specifically applied to the business world. Many argue this movement was sparked by the publication of Paul Hawken's (1993) ground-breaking book *The Ecology of Commerce*. In this best-selling book, Paul Hawken aims to revolutionize the relationship between business and the environment. Sparked, perhaps by Paul Hawken's book or whatever the impetus may have been, organizations around the world began to have a closer look at how Sustainability applied specifically to organizations. After much discussion, many conferences, summits, boardroom meetings, round tables, etc., a refined notion of Sustainability, one that applies specifically to organizational Sustainability, emerged in the early to mid-1990s. Let's take a closer look at what Sustainability means from an organizational perspective.

In the business world, many terms for Sustainability have been used including: Environmental & Social Governance (ESG); "triple bottom line" or People (Social), Planet (Environmental) and Profit (Economic). Today it is widely accepted to simply use the term "Sustainability." Regardless of what one calls it, Sustainability at the micro or firm level is similar to the global view in that it is based

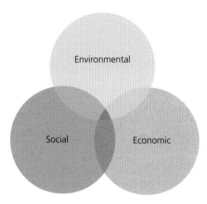

Fig. P.1 Diagram of the three dimensions of Sustainability illustrating how the dimensions intertwine with each other.

upon improving the three basic dimensions of Sustainability: social, environmental, and economic performance (Fig. P.1). More specifically, Sustainability from a business perspective can be defined as continuously improving and balancing social, environmental, and economic performance across the value chain. It is important to note that while Sustainability will look different for every organization, the basic premise and end game remains the same.

Every organization will have different view of what Sustainability means to them based on the aspects that are material to them for each dimension of Sustainability. For example, a car manufacturer may have higher impacts that result from material and energy consumption or the end use of their vehicles whereas a food company may have higher impacts that result from water or agricultural use. This means that the focus or path to Sustainability will be different for every organization but the basic premise and end game of balancing and continually improving social, environmental, and economic performance remains the same. Let's take a closer look at each of the three dimensions.

Social sustainability

While Social Sustainability as defined by both the Brundtland report and The Natural Step focuses on meeting the basic human needs of the people within the system, Social Sustainability at the firm or organizational level is viewed a little differently in that it focuses on what the organization can do to contribute to Sustainability of the larger macro system. Social Sustainability at the firm level strives to ensure that all direct and indirect activities throughout the value chain at the very least have an overall net neutral impact on society with aspirations towards having a net positive impact. This idea can seem rather overwhelming to wrap one's head around; it helps to break it down into the major social impact areas:

People (workplace): Refers to an organization's employees and physical workplace. Many organizations have long focused on health and safety, employee engagement, performance management systems, etc. However, the aspect of people under the dimension of *Social Sustainability moves beyond traditional people focused initiatives such as a safe and healthy workplace and employee benefit programs to deliver an environment which enriches both the personal and professional lives of its people*. This may include things such as specific skill and knowledge development, volunteering opportunities, flexible work schedules, telecommuting, and so on. Many of the world's most successful business leaders such as Sir Richard Branson, the founders of Google, and Mark Zuckerburg among many others have long preached how their people are one of the single greatest catalysts for success.

Partners: Refers to partners across the value chain including but not limited to suppliers, customers, and other key stakeholders such as Government, NGOs, banks, industry associations, and even animals. Like the area of people mentioned earlier, when it comes to partners, organizations have long focused on developing relationships with partners but this has been focused more towards downstream partners such as customers vs upstream partners such as suppliers. Additionally, the focus has been more around managing risk vs maximizing opportunities while also mitigating risk. The aspiration here is to *move beyond traditional partner relationship to develop long-lasting, mutually beneficial relationships to become the supplier of choice for downstream partners and the customer of choice for upstream partners*. Achieving success in an increasingly competitive global marketplace requires an organization to have strong relationships with all of its partners. The "Business Case" section of this primer will dive into the benefits in more detail but in today's marketplace, fostering relationships with suppliers, government, NGOs, etc., can be just as important if not more important than, say, a focus on retaining customers.

Community: Refers to any and all of the communities or parts of society that an organization directly or indirectly impacts. This includes but is not limited to the physical communities where the organization operates or has products/services available, the people that consume or are touched by their products/services, as well as the greater society. Many times this also refers to an organization's philanthropic efforts. While many organizations have long managed community impact, the focus again has been more on managing risk. The aspiration here is to *positively impact all communities or aspects of a society that the organization directly or indirectly touches.* In today's society, communities are playing an increasingly larger role with regards to an organization's social licence to operate. It is becoming ever more common for communities to prevent or restrict an organization's ability to operate or expand due to their negative impacts on the community. Focusing on positively impacting the community will not only help mitigate negative backlash and secure a licence to operate, it will also help develop new customers, drive innovation, and build brand equity. Many organizations with a focus on the community piece of Social Sustainability are finding innovative ways to maximize benefit. The concept of "Creating Shared Value" was coined by Michael Porter from Harvard and is an elegant example of what is trying to be achieved here. Also known as "Strategic Philanthropy", the concept is fairly straightforward. Create value for the community while at the same time creating value for the organization. A great example of this comes from Facebook. Facebook has a large philanthropic strategy in place to provide internet access to areas of the world that – due to lack of wealth and infrastructure – do not have access to the internet; amazingly this number is quite large, according to Facebook two-thirds of the world does not have access to the internet. They call this initiative "Internet. org" and aim to connect the remaining two-thirds of the world to the internet. Think about this; what an amazing benefit and value for these two-thirds of society but at the same time look at the value it brings to Facebook. The more people connected to the internet, the more Facebook users there will be, in fact with a global population around 7 billion, this means a possible 4 billion+ more Facebook users – what would this do to their performance?

Continually improving in each of these 3 areas will increasingly reduce an organization's social impact while delivering tremendous value to the organization – value that previously was left on the table but because of an understanding and focus on Sustainability is now being realized. As organizations continue along this trajectory, impacts are reduced and benefits increase resulting in enhanced organizational performance while at the same time moving towards having a positive impact on society.

So while the areas that Social Sustainability focuses on at the firm level are not necessarily new focus areas per se, the end goal or aspiration for the focus changes from one of minimizing impact and managing risk to one of maximising benefits and driving organizational performance. It provides a lens which organizations can look through to ensure they haven't missed anything in this area while providing a view of what the future can look like.

Environmental sustainability

Commonly referred to as going "green," *Environmental Sustainability strives to ensure that all of an organization's direct and indirect activities throughout the value chain have, at a minimum, a neutral impact on the environment with aspirations towards having a net positive impact.* Referencing The Natural Step's definition of Sustainability provides clarity in better understanding what this means.

In order to have a neutral impact on the environment, an organization must fully and completely satisfy the first three of the four system conditions provided by TNS. This means firstly that organizations must not extract resources form the Earth's crust faster than they can be regenerated. In other words, organizations must power operations using only renewable energy not fossil fuels and stop pulling resources such as copper from the ground or find ways to reuse them over and over again once extracted. This requires a fundamental shift in design philosophy whereby industry moves from a linear philosophy where virgin raw materials are designed to end up in the landfill to a circular philosophy whereby recyclable materials are designed to come back to the producer and be born again into a new offering.

Secondly, organizations must not discharge pollutants and toxins into the environment (including water) faster than the Earth can clean or deal with them and not discharge toxins or pollutants into the environment that the Earth has no capability to clean or deal with. In other words, organizations must phase out the use of toxins the Earth cannot deal with such as certain pesticides and fire retardants while ensuring ongoing discharge of any other pollutants is done at a rate the environment can deal with. Due to heavy regulation and the increased availability of greener chemicals, many organizations are already moving down this path but there is certainly room for improvement and a much wider adoption of this concept.

Finally, organizations must not destroy or harvest surface level resources faster than they can be regenerated. In other words organizations must use surface level resources such as trees, wildlife, water, plants, etc., at a rate that can be continued indefinitely. Again, this requires a fundamental shift in thinking towards operating practices that allow time for regeneration. Many organizations are already on this path with such practices as sustainably harvested seafood, sustainable agriculture practices, use of recycled materials vs virgin materials, and so on.

It is not until these three system conditions have been fully and completely satisfied across the entire value chain that an organization is truly environmentally sustainable or "green." The focus or path for satisfying these three system conditions will be different for each organization because different resources are being extracted, different pollutants are being discharged, and different surface level resources are being harvested or destroyed based on the products or services offered by the organization. Fully and completely satisfying these three system conditions across the entire value chain is extremely challenging and can take decades to achieve. In fact, there are no (or very few) organizations in the world today that are truly green. Understanding the true meaning of being green is important not only to determine the path forward but also to avoid a false sense of success that could deteriorate the brand and subtract from efforts that would bring value and benefit to the organization

Economic sustainability

Contrary to popular belief, Economic Sustainability is about more than managing costs and next quarters' financial statements. It moves beyond traditional fiscal control and ***strives to ensure that all of a firm's direct or indirect activities throughout the value chain contribute towards sustained demand and supply of products and/or services.*** In other words, engaging in activities that will help ensure the doors stay open for 10, 20, even 50 years from now.

Once again, this idea seems formidable but Economic Sustainability does provide some guidance in the form of high-level concepts to focus on. While not an exhaustive list, the following concepts provide a good starting point for driving long-term economic performance.

Innovation: Refers to the continuous process of inventing, developing, modifying and evolving new or existing products, services and processes. Through innovation, organizations continually stimulate a fresh appetite for their products and services contributing to sustained economic performance. There are many great modern day examples of innovation such as Yoplait's GoGurt Yogurt tubes that makes eating Yogurt more convenient (do not need a spoon) or Evian's Smart Drop which allows customers to order more water at the touch of a button that is attached to a magnet on the fridge. This innovation addressed an issue of lugging heavy water from the grocery store that was identified through customer feedback as a barrier for greater consumption. Other innovations such as Apple's iPod and iPad or Dyson vacuums also illustrate the importance and benefits of innovation.

Agility: Refers to the ability of organizations to quickly and effectively adapt and respond to rapidly changing market conditions. Failure to respond to shifting conditions in the marketplace may cause the organization and/or its offerings to become unfitting or irrelevant. Having the ability to quickly adapt to changing market conditions will support sustained demand for products and/or services. Again, there are some great examples of organizations that are successful in responding to market conditions and fruitfully benefiting from the adaptation. PepsiCo for example began to recognize a shift in the marketplace towards healthier snacks and beverages leading to the development and overwhelming success of a new juice line called TROP 50 which contains 50% less sugar and calories than traditional Tropicana products. On the flip side there are many organizations who failed to develop this skill of agility and no longer exist. Blockbuster, for example, is a great example of the antithesis of Economic Sustainability. Due to lack of innovation and agility, the organization was blindsided by online streaming services such as Netflix.

Resilience: Is the ability to recover after a disturbance. In nature, disturbance may come in the form of a forest fire, a storm, or a rampant disease. In business, disturbance may come in the form of a natural disaster affecting supply of materials, terrorist attacks, resource scarcity, economic turmoil, health pandemics, and the list goes on. When business was simpler, resources and markets were abundant and limitations were few, conventional thinking served organizations well.

As the world becomes more unpredictable, complex, dynamic, and disturbances increase in number, impact, and volatility, organizations must have the ability to respond in order to survive.

While there is no secret formula for Economic Sustainability, addressing the areas discussed earlier by putting systems and strategies in place to support them will certainly help to build a solid foundation for future success.

Clearly, Sustainability is about more than saving energy or "going green." The environmental dimension is only one piece of the puzzle. The challenge (or opportunity) that Sustainability presents is effectively balancing performance within all three dimensions; environmental, social, and economic. For example, an organization can be truly green but if there are no orders coming in the door due to lack of innovation or there are disruptions in the supply chain from ignoring partners or the organization fails to retain and attract talent, they become unbalanced and thus unsustainable over the longer term. Think of a three-legged stool. If one of the legs is shorter or longer than the others and any pressure is applied, the stool will collapse. Sustainability can be viewed in a similar light in that if the focus across the dimensions is not balanced and any pressure is placed on the organization, it could collapse just like a wobbly three-legged stool.

The power of Sustainability is not necessarily in each of its components but rather the sum of its components. In other words, many of the individual concepts presented by Sustainability are nothing new. Organizations have been focusing on developing their people, giving back to communities, developing partner relationships, reducing environmental impacts, and driving innovation for decades. Unfortunately, most organizations have an incomplete or unbalanced approach to Sustainability in that they ignore, either in part or full, one or more of the dimensions or have an unbalanced approach across the dimensions. So an organization that has a strong approach to its people may ignore or have less focus on the environment or a company with a strong focus on the environment, its people, and innovation may not be focusing on partners or the community. Whenever an organization fails to have a complete a balanced approach to Sustainability, they are leaving opportunity on the table. The beauty of Sustainability is that it provides an approach to business that encompasses all the variables that drive business performance. Think about Sustainability as super-powered glasses. When an organization puts on the Sustainability lens, they can see the missing pieces to their success.

Business case for sustainability

Sustainability is not some sort of lefty, tree-hugger movement with an altruistic view of the future; it is in plain and simple terms – *a better way to make a bigger profit*. Over the past two decades, organizations across the globe have shown how a focus on Sustainability drives performance and profit. Organizations such as Walmart, GE, and Unilever have made Sustainability part of their strategy, not because it makes them feel warm and fuzzy inside but because it elevates their performance and drives profitability. Understanding the business case is an important step in the journey, so let's look at the main categories to the business case:

Direct cost savings

It doesn't matter whether one believes in things like climate change, resource scarcity, treating employees fairly or giving back to the community at the end of the day if an organization uses less energy, consumes less water, retains talent, can quickly react to business disruptions, etc., they spend less money. Plain and simple.

Now, some folks may argue that green costs more. This is a myth. The argument that green costs more is based on a short-sighted and incomplete view of cost. The basis of the "green costs more" argument is that "green" solutions such as energy efficient equipment cost more to purchase. While this may be true in some cases, this argument only looks at first costs or the initial purchase price. When evaluating the cost of a solution, one needs to look at the total cost of ownership. Total cost of ownership includes not only the purchase price but also includes operating costs over the life of the equipment, maintenance requirements, life expectancy, and disposal costs. Once the total cost of ownership is taken into consideration, the "green" option usually has a lower total cost of ownership. Take lighting for example. The upfront cost to purchasing more efficient lighting such as T8 or T5 bulbs to replace metal halides or traditional lighting definitely has a higher purchase price or "first cost." However, once one takes into consideration the reduced energy costs to operate the lights over the lifetime, extended life of the bulbs, reduced maintenance costs, improved illumination, and so on, many times the payback for the added "first cost" is typically two years or less without any incentives, of which there are many. When simply looking at first cost, sure, green may cost more; however, looking at the total cost of ownership, green solutions cost less.

Traditional business models have typically viewed green-related expenses such as energy and waste disposal as fixed costs to doing business. This is completely false. "Green" related expenses such as energy, water, and waste disposal are really variable costs and need to be managed as such. If one does not want to call it green, call it variable cost management because that is inherently what it is.

Increased revenues

As more and more individuals and businesses realize the importance of Sustainability, they are demanding partners who share the same values. The Sustainability movement can be compared to the quality movement in the 1980s. Initially businesses were not very focused on developing quality management programs certified to internationally recognized standards. As the larger corporations and government organizations started to demand quality management systems as a prerequisite for doing business, organizations across the globe began developing, implementing, and certifying quality management systems to retain current customers and attract new ones. Leaders in this area were rewarded with increased business and laggards were eventually forced to follow suit. Today, having a quality management system is simply a ticket to the game, an expectation.

Sustainability is taking a similar route in that more and more individuals and businesses are looking for, and in more cases demanding, Sustainability systems and strategies be in place. Take Walmart for example; Sustainability performance is now a performance criteria for all suppliers in addition to traditional criteria such as cost, quality, and delivery. Additionally, a Sustainability Performance Index for products based on impacts throughout the life cycle is being rolled out. Based on the performance under this index, a score is assigned to each product and buyers have incentives built into their performance management systems for procuring more sustainable products. More and more of these types of programs are being rolled out by other retailers and their suppliers, and many other industries and sectors are following suit. The result is that leaders in this area are being rewarded through retention of current customers and stealing market share from competitors who are laggards in this area. However, the window of opportunity to leverage and capitalize on this movement is beginning to close. Within the next 5–10 years, a strong commitment to Sustainability will mimic the "quality" movement in that it will be an expectation or prerequisite for doing business. There are many studies that clearly show this movement is taking place.

Employee engagement, retention, and attraction

With more people realizing the importance of Sustainability, a rising number of employees are seeking out employers who are committed to Sustainability. This is especially true with younger generations. These folks have grown up learning about the importance of environmental and social responsibility and are hungry not just for a paycheque and some creativity in their job, but for employers who also believe in and practice Sustainability.

Organizations committed to continuously improving their social, environmental, and economic performance have experienced the power this brings with both retaining and attracting new employees. Even with recent economic turmoil around the globe, many individuals are choosing to seek out sustainably minded organizations. A recent article from the *Harvard Business Review* stated that a new study showed that a survey of 750 employees from across China, Brazil, the UK, and the USA showed that two-thirds of respondents stated that sustainable business practices are extremely important to them.(2) As any successful business leader will tell you, having the ability to retain and attract top talent is critical to success and Sustainability has proven to be an effective means for achieving this.

Risk management

While cost savings and increased revenues have historically been the main driver for engaging in Sustainability, many organizations are beginning to look at it from a different angle. A number of organizations are starting to investigate the risks associated with not having a focus on Sustainability and finding out that the risks are more serious than previously understood.

As one can imagine, there are a number of risks associated with a lack of focus around Sustainability. To begin with, the screws around environmental and social legislation are not getting any looser. Governments around the world are elevating reporting requirements, lowering allowable thresholds, introducing carbon and other environmental penalty schemes, while generally increasing accountability for overall social and environmental impacts. A lack of attention towards key Sustainability issues could find many organizations progressively out of compliance and subject to unforeseen penalties, fines, etc. A strong focus on Sustainability will not only help ensure compliance today but tomorrow as well.

Key Sustainability trends such as social responsibility, human rights, climate change, more frequent or severe weather events, and water scarcity are becoming ever more prevalent. Organizations who rely on the Earth's climate and resources to directly produce and supply raw materials or are subject to a high level of consumer scrutiny, such as those in the food sector, are especially vulnerable to being impacted by these trends. Many organizations, such as Unilever, have realized that even without any direct business benefits such as cost savings, just the risks of not doing anything provide enough incentive to become engaged in Sustainability.

Increasing shareholder value

Investors across the world have begun to realize that organizations driven by Sustainability are outperforming those who are not. Whether it be through improved bottom line or top line performance, retention and attraction of top talent, or effectively managing risk, the results are clear: those who focus on Sustainability outperform those who do not.

In addition to all of the other benefits of focusing on Sustainability, increasing shareholder value and attracting new investors is the icing on the cake. It is also important to note that there are a number of key organizations and agencies working to make it a requirement for publicly traded organizations to monetize and report environmental and social performance in addition to fiscal performance. This will certainly drive many of the large organizations to increase their focus and performance around Sustainability.

At the end of the day, spending less money, improving market share, retaining and attracting talent, managing risk, and attracting investors will all result in greater profits and overall performance both now and in the future. Sustainability is not just the right thing to do but the smart thing to do.

Introduction – Setting the Stage

Simplicity is the ultimate sophistication.
Leonardo Da Vinci

In the pages that follow, you will come to learn and understand a powerful system for maximizing organizational performance. This is not something that can be learned in a University classroom or by completing an MBA. It comes from over 15 years of experience as an industry practitioner, a professor, consultant, and author. Furthermore, it comes from getting inside hundreds of organizations coupled with a passion for studying what separates the great companies from the mediocre ones. It comes from a drive to find a better way.

It has involved countless hours of boardroom discussions with high-powered executives and small business owners, endless days spent with front line leaders, too much time spent in quiet reflection on airplanes travelling to and from different organizations, participating in numerous conferences, and learning from a myriad of mistakes. All with the end goal of developing a simple but powerful process for developing a Strategic Planning system that maximizes organizational performance and enables agile responses to rapidly changing markets while positively impacting society and the environment.

Shouldn't be too hard, right? After all, people are printing functioning human organs, building elevators to space, and bringing back extinct animals from a sliver of DNA. The truth is that a number of organizations have already figured it out, so in a sense, all one has to do is look. These organizations may not say or even know that they have figured it out but they have or are pretty close to it. It is amazing how successful organizations all share similar traits, characteristics, approaches, philosophies, etc., to strategy development and execution. Even organizations in completely different industries or with completely different types of leaders and cultures, the successful ones still share an astonishing amount of similarities when it comes to things like strategic planning and execution, alignment, culture development, engagement, priorities, management traits, philosophies, etc. This makes sense since a lot of companies face the same challenges and opportunities – even though, for some reason, a lot of folks do not want to admit this. The trick is putting all the pieces of the puzzle together in a Strategic Planning and Execution system that is simple but effective. This book does that.

Strategic Planning is nothing new, there are countless models available, some very effective, and others not so much. Furthermore, most organizations have some sort of strategic plan in place, some more formalized, some less formalized, some very successful, and others an epic failure. Reinventing the wheel on strategic planning is boring – that is not what this book is about. This book is about disrupting the current approaches and shifting the paradigm on strategic planning to reflect reality and it does that in a number of ways. One way is that it places a purpose beyond profit at the center of the strategy – more on that in the next chapter. Moreover, it also provides a system of strategic planning that embraces change instead of resisting it by incorporating the spirit of Sustainability to drive triple bottom line performance.

Over the past decade, it is very clear that there has been a major paradigm shift taking place in business – it's called Sustainability. Some people equate the paradigm shift of Sustainability to that of the internet – in other words, it is disruptive. This is not about Sustainability in the narrow sense of the environment or "going green," Sustainability is much, much broader than that. (See the "Sustainability primer" for a more detailed look at the broader concept of Sustainability.) Like the internet, many organizations have embraced Sustainability; sure, there are leaders and laggards but over the past decade a majority of organizations from the giants such as Coca-Cola, General Electric, and Unilever to the small family business have adopted Sustainability in some form or fashion. Why? Because it makes sense. It's a better way to make a bigger profit. The rub is that for the most part, current strategic planning models were developed prior to this paradigm shift taking place and therefore do not properly address it. The exclusion of Sustainability at the strategic level has caused a problem for many organizations as they struggle to realize their full potential.

The diagnosis of the problem is pretty simple if one takes the time to study it. What happened is that organizations have been plugging along with their current business strategy (regardless of whether it is a sound one or not) and then come across this thing called Sustainability. Folks get excited about this movement and the smart organizations recognized early on the power of Sustainability to drive organizational performance and embraced it. Recognizing the power of Sustainability to enhance performance, many organizations eventually recognized the need to formalize an approach and developed a sustainability strategy. Sounds logical right? Well, herein lies the problem. This approach results in the organization now having two strategies. One strategy for the business and one for Sustainability – two separate and siloed strategies.

Over time, this caused many organizations to lose focus on Sustainability to where it has almost become a side bar –a nice to do when there is extra time and resources and everything is going well – big mistake! The smart organizations recognized the problem and began looking for ways to integrate the two strategies and achieve what consultants love to call "Embedding Sustainability into the DNA of the Organization" – more on that later. The issue here is that because the existing strategic planning models did not formally address Sustainability, many of the organizations simply tacked on the overall notion of Sustainability to the business strategy. This has been done in many different ways, some organizations added it as a "Pillar" or high-level focus area, others added it to their core values, others simply added it as a line item under an existing pillar or focus area. The result has been that many organizations, in an attempt to place greater focus on Sustainability, failed to properly integrate sustainability into the overall strategy.

The model presented in this book deals with the lack of a sustainable strategic planning system and provides a proven, dynamic, and successful solution to integrate sustainability into the overall business strategy. Instead of building a "Business Strategy" and then building a "Sustainability Strategy" the model integrates the two together to develop a strategy that is Sustainable as illustrated in Fig. I.1. Having an overall purpose beyond profit for developing a strategy in the first place, changes the game even more.

The question then becomes how to do this. It starts with a framework based on a purpose beyond profit. The purpose is supported by a vision of the future which is achieved through focus areas, referred to as pillars. These strategic pillars are the high-level elements that are critical to organizational success. Ensuring that within these pillars all

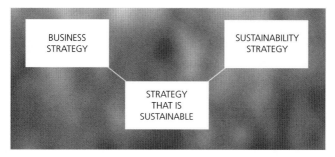

Fig. I.1 Visual illustration of integrating two separate strategies to build a strategy that is sustainable.

aspects of sustainability are covered, is what drives triple bottom line performance. Pointing all the pillars in the same direction, towards a future desired state, creates alignment.

Many organizations today struggle in realizing their full potential not just simply because Sustainability is not properly integrated into the overall strategy or there is no greater purpose beyond profit for their existence. Many times, it is because the entire organization is not aligned and focused on a common vision of the future. The result, is organizations implement a bunch of initiatives and have people running around working on things that do not add value. In other words, organizations are working on a whole bunch of "stuff" but due to the lack of alignment and focus, it is not the "right stuff."

For example, when it comes specifically to Sustainability, many organizations today have some sort of environmental initiatives in place. However, without clearly understanding what environmental aspects and impacts are really important to the longer term direction, organizations come up with some generic project that is popular today, like a "recycling" program. That's great but what if waste is not really a big issue or cost for the organization relative to say energy or water conservation or shaving materials out of a design? The answer is simple; it is the wrong initiative to be working on. Even though it is nice to do, the efforts do not create real value and are pointed in the wrong direction since they do not align with the longer term picture of the future.

Likewise, many organizations today have some sort of philanthropic effort in place – one aspect of Sustainability. This is great; however, if the philanthropic efforts only create value for the recipient and do not create real value for the organization beyond good PR, these efforts could also be pointed in the wrong direction as they may not support the longer term picture. Instead, they should create value for both society and the organization, a concept known as "Shared Value" and coined by Michael Porter from Harvard. Obviously these are simple examples and not all organizations make these mistakes but it is amazing how many do.

This misalignment also rings true for things that are not necessarily considered "Sustainability" related. Take for example the almighty focus on the customer. Organizations spend a lot of time and resources trying to make customers happy, so they are more loyal and buy more products leading to increased revenues. To drive this, many organizations like to focus on delivering "superior customer service." A typical scenario is when A senior leader notices either through a survey or social media or perhaps in discussion with others that there are complaints about customer service. Maybe wait times when calling in for support are too long or customers are not happy with the results that came from calling the support line or something of that nature. So, what ensues is a big initiative to transform customer service – calls must be answered within 3 minutes or responses must be provided within 24 hours or whatever it may be. The thing is, what if the issue isn't actually the customer service but something else? Maybe the product or service is poor to begin with and therefore, no matter how fast a call is answered or how great the customer service, the customer is not going to be happy because the product was no good or did not deliver as promised. In this case, the same thing is happening; efforts are pointed in the wrong direction.

Obviously these are simple examples and not every organization operates this way but it is frightening how many organizations have not even identified the factors critical to success or the direction they want to head, never mind making sure that all activities are aligned to a common picture of success.

In order to ensure efforts are pointed in the right direction, organizations must have a clear vision of the future and be laser focused on areas that are critical to successfully making the vision a reality. Beyond this, organizations must also ensure that their efforts include all dimensions of sustainability in order to optimize performance.

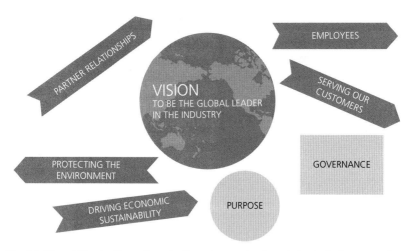

Fig. I.2 Visual illustration of a misaligned organizational strategic framework with aspects of Sustainability missing.

Fig. I.2 illustrates what a typical misaligned organization looks like. The organization is "busy" working on a bunch of stuff but it is not the right stuff because efforts are pointed in the wrong direction – they are not aligned with the long-term "vision" of the organization. Additionally, because Sustainability has not been properly integrated into the overall strategy, there are pieces missing. Where is the "community" or philanthropic aspect of Sustainability? If this is a focus for the organization, it is not clear or it has been hidden somewhere. Furthermore, why does the organization exist in the first place? There seems to be no purpose beyond profit for existing or the purpose has been pushed aside and is no longer a central component to driving the overall strategy.

The framework shown in Fig. I.3, which is the foundation for the strategic planning system presented in this book, addresses a number of the issues with the current approaches to strategic planning. First, it corrects the misalignment by requiring a clear Vision of the future that is supported by a set of critical success factors (aka "Pillars" such as "Employees" or "Partner Relationships" in Fig 1.3) which keep the organization focused. Furthermore, it addresses the failure to properly integrate sustainability by ensuring all aspects of Sustainability are covered across the pillars. Finally, having a

Fig. I.3 Visual illustration of a properly aligned organizational strategic framework that includes all dimensions of Sustainability with a purpose beyond profit maximization.

purpose beyond profit as the central motivator, changes the game of Strategic Planning.

One can clearly see the difference between the before and after. In the Fig. I.3, the organization has identified the focus areas critical to success in the form of pillars, which contain all dimensions of sustainability and are clearly aligned to the long-term vision of the organization to fulfill a purpose beyond profit maximization.

It is great to have a theoretical framework and nice diagrams to show what things should look like. But how do organizations actually use this framework to build a Strategic Plan that is aligned and successfully integrates all aspects of sustainability while having a greater purpose? Furthermore, how do organizations move from this framework to build a Strategic Plan that will naturally generate the activities required to realize the future desired state while being able to measure performance along the way? The answers to these questions are presented in this book.

In the chapters that follow you will learn how to build a Strategic Plan that is aligned, focused, and integrates all aspects of sustainability while deliberately fulfilling a greater purpose. It is based on over 15 years of experience that has culminated in the development of a step-by-step process that has been fine-tuned through trial and error,

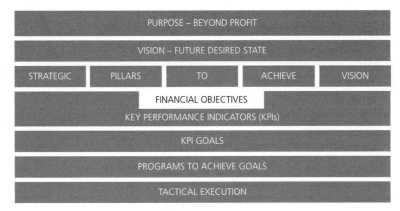

Fig. I.4 Visual illustration of the process for developing a strategy that is Sustainable.

with real companies over the past 5 years. For organizations that have adopted this approach, the results are incredible; double digit margin improvements, increased revenues, greater employee engagement, enhanced risk mitigation, and elevated stakeholder value.

The process for developing a strategy that is sustainable is illustrated in Fig. I.4.

It all starts with finding the purpose – beyond profit, for the organization. This is the reason why the organization exists. As will be seen, purpose serves as a great motivator for employees and other stakeholders while also providing a north star that will guide organizations along their journey. With a purpose in place, a vision of the future state will be developed. Achievement of this future state is required in order to fulfill the purpose and serves as the point of alignment for all activities across the organization. The next step is where the "magic happens." A set of strategic pillars are developed that will provide the focus areas that are critical to organizational success and required to achieve the vision. Collectively, these pillars will include all aspects of Sustainability, allowing Sustainability to be embedded into the overall strategy and laying the groundwork for building a strategy that is Sustainable. These pillars will help keep the organization focused and aligned.

Once the pillars have been developed, a set of financial business objectives are developed. Following this, a set of Key Performance Indicators (KPIs) aligned to the pillars will be developed to measure

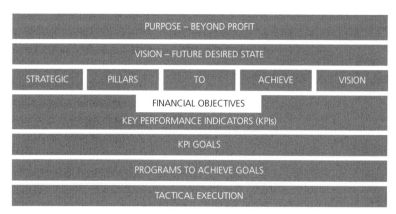

Fig. I.5 Strategic Planning Process Diagram.

overall organizational performance. These KPIs are critical to the overall strategy since they will be used to develop clear organizational goals. The KPIs together with the goals create a mechanism for building accountability, driving action, and achieving the financial objectives.

With clear goals set, the next step is to leverage the foundation that has been built to develop the programs and initiatives that will be put in place to achieve the goals that have been set. These programs and initiatives will drive all the improvement activities that employees will be working on.

The final step in the process focuses on how to execute the strategy that has been developed. An additional appendix is also provided that presents a suite of practical tools and resources to support the overall design and execution of the strategic plan.

This can be a lot to digest, so to recap, let's have another look at the process for designing a strategy that is sustainable (Fig. I.5):

In summary, the system produces a strategy where all the "stuff" employees are working on (Programs in Fig. I.5) adds value by contributing to achieving the Goals that will improve the KPIs which drive the pillars to achieve the Vision in pursuit of fulfilling the Purpose. Profit is still part of the equation but it is an intended by-product of fulfilling the Purpose in a sustainable way.

Not only will the strategy be sustainable and have a purpose, everything is aligned up and down, over and across. It will accelerate the velocity at which organizations improve performance and realize profit while at the same time doing it in a way that positively impacts the environment and society. By following the process detailed throughout this book, organizations will have a strategy that is sustainable, where there is purpose to making a profit, and profit is realized on purpose – the organization will be Purposely Profitable.

The nature of the model and the way the process is laid out, also makes this an agile approach to strategic planning. An organization can adapt the details of the strategy – such as goals or even the KPIs or Pillars – to respond to market changes and make midcourse corrections without compromising the long-term direction of the strategy.

Beyond providing a proven approach to developing a strategy that is sustainable, the value of this book is that, unlike many of the business books on the market today, it goes beyond laying out a theoretical model to show not only what needs to be done but also how to actually get it done. It is designed to coach to competency on the process for developing a strategy that is sustainable. It's like having a high-powered consultant at your fingertips, at a fraction of the cost. It will take some work to develop the strategy but the process is proven, it is dynamic, it is practical, and most importantly, it works.

In order to realize the greatest value from this book, follow the chapters as they are laid out. It has been organized according to the steps in the process. Order is important, so each step must be completed prior to starting the next and each one is just as important as the others. It is suggested to read through the book in its entirety before actually starting the strategic planning process. This will provide an understanding of the process and help prepare for carrying out the process. The book can then be referred to as needed throughout the process to provide guidance and support.

CHAPTER 1

Finding Purpose

It is not enough to be industrious; so are the ants.
What are you industrious about?

Henry David Thoreau

Google, Whole Foods, Patagonia, Tesco PLC, Walt Disney ... what do these ultra-successful organizations all have in common? Purpose. Each has a clearly defined, easily understood, and meaningful purpose – beyond profit, at the heart of everything they do.

An abstract concept that conjures up images of wise elders philosophizing about the meaning of life, purpose can be a hard thing to wrap one's head around. The purpose of an organization is a little bit different from the purpose of life though. Clearly, organizational purpose hovers at a different level from the purpose of life but it does provide similar value. Fortunately, it is a lot easier to figure out as well.

At its core, organizational purpose is in fact a rather basic concept. It defines why the organization exists. It answers why the organization's work is important and the contribution it makes to society, environment, and the economy. It does not answer what the organization does, but rather why it does it. It provides the inspiration for innovative product and service offerings, it is the impetus that drives all strategies, decisions, and activities, and it is a central component of culture.

Purposely Profitable: Embedding Sustainability into the DNA of Food Processing and Other Businesses, First Edition. Brett Wills.
© 2016 John Wiley & Sons, Ltd. Published 2016 by John Wiley & Sons, Ltd.

If, therefore, organizational purpose defines why the organization exists and assuming that fulfilling the purpose will provide a benefit, then *a purpose-driven organization aims to maximize benefit over profit.* Sure, organizations must be profitable to be sustainable but profit maximization is the result of strategically fulfilling the organizations' purpose.

To better understand this concept, it helps to look at some examples of organizational purpose from some of the world's leading organizations (Game Changers 500).

 Organize the world's information and make it universally accessible and useful.

 Produce quality apparel to inspire environmental solutions.

 Helping support the health, well-being, and healing of both people – customers, team members, and business organizations in general – and the planet.

 Making what matters better, together.

 To make people happy.

Clearly, purpose will vary from organization to organization, but at the core it defines why the organization exists and what benefit(s) it provides. These may sound like simple questions but in their simplicity lies tremendous value for the organization and its stakeholders.

It is also important to note that the purpose of an organization differs from its vision and its mission – more on these in later chapters but essentially, the vision is a future desired state achieved by carrying out its mission on the journey to fulfilling a purpose. In other words, the mission describes what is required or what the organization will focus their efforts on in order to achieve the vision that will fulfill the purpose.

1.1 Why a purpose?

With a better understanding of what is meant by organizational purpose, the natural questions to ask are: What benefits does it provide? Can organization be successful without having a clear and meaningful purpose? Sure. Organizations have been and will continue to be successful without having a purpose, but a paradigm shift is taking place in business. The rules for succeeding – for creating strategic advantage – have fundamentally changed. Extensive research has proven that the traditional *shareholder-biased* business model, where profit is the purpose and pursued at all cost, is an inferior generator of value.

On the contrary, a new *stakeholder-based* business model where greater profit is earned as a natural by-product of fulfilling a greater purpose – that benefits all stakeholders – is the most effective way to maximize performance and manage growth over the long term.

According to extensive research from the consulting magnate Deloitte, when Chief Executive Officers (CEOs) and other business leaders convey a strong sense of purpose behind their strategy, they help propel the organization toward multiple, positive revenue-influencing, cost-reducing outcomes. Businesses which rank high on purpose-based qualities are better positioned for brand strength and customer loyalty, revenue growth, and agility amid disruptive market changes. Their employees are more engaged, more likely to buy into corporate strategies, and place greater trust in the capabilities of their leadership teams. Shareholders also express increased confidence in the ability of these organizations to meet performance and earnings targets (Deloitte Development LLC, 2014).

Why is this? Why does something as seemingly simple as a purpose result in such positive outcomes? When an organization has a clearly defined and meaningful purpose, it does a few things. First, it provides all stakeholders with a clear reason for why the organization exists. Second, these stakeholders are really just people at the end of the day and most people have an itch to support or contribute to a greater good that generates benefit for something they care about. For employees, it gives meaning to mundane, day-to-day activities and provides something to get fired up and passionate about. Through their work, employees can make a difference and be part of a meaningful legacy. When an organization's purpose is meaningful to an employee, that person feels a connection to work that is not only rational, it is also emotional. For customers, shareholders, suppliers, and other key stakeholders, it provides a sense of supporting a greater cause, creating an emotional link to an otherwise emotionless transaction. In other words, it checks the proverbial "box" most people have, to do good.

Let's not start reciting lines of "Kumbuya" here. People's decisions and actions are dependent on many factors. Nonetheless, the results are clear. Organizations with a strong sense of purpose are clearly benefiting from it and organizations without a clearly defined purpose may be showing up to a playoff game and voluntarily sitting out the star player. It's still possible to win, but with the fierce competition that comes with the globalized market, voluntarily taking the chance would be gambling on success.

1.2 Finding purpose and developing a purpose statement

Unfortunately, there is no scientific formula for developing a purpose; this is something that needs to grow organically. The good news is that most organizations have (or had) a purpose and it has been lost along the way or pushed aside. The challenge for most organizations is in re-finding this purpose and being able to clearly articulate it in what is known as a purpose statement. What follows is a simple, straightforward process for identifying organizational purpose and articulating this purpose through a clear purpose statement – not dissimilar to the examples provided earlier. Simple but powerful, the process uses key words to spur the development of a purpose statement which clearly articulates the purpose of an organization.

Before working through each step in the process, it helps to better understand the characteristics of a strong purpose. Organizational purpose should be:

- **Inspirational:** Fires people up and gives them something to be passionate about.

- **Meaningful:** Provides true value to stakeholders and/or society and/or the environment and/or the greater economy.
- **Relevant:** Aligns with the organization's core competencies, products, and/or services.
- **Authentic:** Not some marketing ploy to temporarily increase sales but a genuine passion for providing benefit.

With the characteristics of a strong and meaningful purpose in mind, let's look at a simple exercise to identify and clearly articulate organizational purpose. When working through this exercise and the exercises for the other steps in the Strategic Planning process, it is important to include the right people who will add value while being careful to not hinder progress by having too many people involved. Furthermore, in order to provide greater understanding of the exercises and overall process, a fictional company will be used as an example. To provide some context, this fictional company will be a mid-sized food processor located in North America. Let's call this fictional company Smithville Foods (any similarities to an existing organization is purely a coincidence).

Step 1: Articulating the purpose

This first step focuses on starting to articulate the essence or spirit of why the organization exists or, in other words, the purpose. To help do this, there are some key questions that can be asked. Let's work through these questions using Smithville Foods as an example. A quick note here – feedback on the initial draft of the purpose statement can be attained from key internal and external stakeholders but the statement should be developed internally first. Stakeholder engagement is important and will come into play later but it is not up to the stakeholders to determine the purpose. Here are the questions along with answers from Smithville Foods.

Q:	Why was the organization founded?
A:	To Provide quality food products.

Q:	What problems does the organization solve?
A:	Lack of affordable and healthy food.
A:	Access to fresh food.
A:	Access to natural food.

> Q: What benefit(s) do the products and/or services provide to
>   stakeholders, society, and/or the environment?
> A: Nourish people's health and well-being.
> A: Reduce obesity, diabetes, and other food-related health issues.
> A: Fair access to fresh, quality food.
> A: Positive impacts on the environment throughout the life
>   cycle of the products.

There may be additional questions that come to mind when working through answering these initial questions – if they help with the process, use them. The point of this is to spark ideas that begin to paint a picture of the essence of the organization's purpose, so do not be limited by just the questions listed. Looking at the example answers, it is becoming clearer that the essence of the purpose for Smithville Foods is around natural, healthy, affordable foods that positively impact people's health. While this is not yet a purpose per se, the essence of the purpose is starting to come out and that is what should be happening at this point in the exercise.

Step 2: Crafting a purpose statement

Bringing a purpose to life requires it be communicated. This starts with a purpose statement that describes the purpose of an organization in a few powerful words that inspire, motivate and create emotional attachment with stakeholders.

Here are a few guidelines will help to ensure the statement realizes the greatest impact:

• Use powerful words that are meaningful to employees and stakeholders.
• Be brief in length so people will remember it.
• Be broad in scope to allow for future opportunities and change without having to rewrite the statement.

Using the words in the answers developed in Step 1 and with the essence of the purpose in mind, begin drafting potential statements that can be used to articulate the purpose by stringing the different words together to form a sentence. Clearly this is more art than science and it may help to pull in some of the more creative staff members for this. Continuing with the example of Smithville Foods, a few potential statements could be:

To nourish people through easy access to healthy food.
To combat obesity through affordable and accessible food.
To improve health and well-being through access to fresh produce.
To improve health and well-being.

Do not get too caught up with fine tuning the statements at this point. As long as the statements articulate the purpose clearly, wordsmithing the statement will be part of the next step.

Step 3: Finalizing the purpose statement

With a series of purpose statements drafted, the organization could choose to let key stakeholders such as employees, customers, suppliers, and so on provide feedback on the potential statements. This could be achieved through one-on-one consultation with stakeholders or perhaps through a survey where stakeholders can vote on the statement that speaks loudest to them.

Based on feedback provided around the draft statements – whether internal, external or both – the organization must now decide if there is a particular statement that resonates the loudest.

This may be as simple as choosing one of the statements based on feedback provided or maybe wordsmithing one of the existing draft statements to incorporate feedback or provide greater clarification of organizational purpose. Continuing with Smithville Foods, a finalized purpose statement for them could be:

To improve the health and well-being of current and future generations.

At this point, the organization should have a clear picture of its purpose, which is clearly articulated in a purpose statement like the earlier example. At first, this exercise may seem too simple, but do not be fooled. It is designed that way and it is not nearly as simple as it first seems. While it only takes a few minutes to read through the process, working through the process takes some time as it begins to spur valuable discussions and force people to really think about why the organization exists and how to best articulate it. Often there will be disagreements on not only the purpose itself but also how to articulate it – this is good! This is what needs to happen as it brings people together to hash out differences of opinion and align everyone to a common purpose. Finalizing the statement also takes some time but at the end of it all, it is well worth the effort when a clear purpose for the organization is revealed.

It is important to note here that the power of a purpose does not come from being shared by a select group of executives or from hanging on a wall in a boardroom, it comes from all stakeholders clearly understanding and buying into the purpose. This can only be done through clear and constant communication of the purpose using the purpose statement – more around this in the chapter on execution later in the book.

At this point in the process, a clear purpose should have been revealed along with a statement developed that clearly articulates this purpose. The first step in the strategic planning process for developing a strategy that is sustainable is now complete. What follows is a progress indicator that will be found at the end of each chapter. It will show which steps in the strategic planning process have been completed.

Progress indicator

PURPOSE – BEYOND PROFIT				
VISION – FUTURE DESIRED STATE				
STRATEGIC	PILLARS	TO	ACHIEVE	VISION
FINANCIAL OBJECTIVES				
KEY PERFORMANCE INDICATORS (KPIs)				
KPI GOALS				
PROGRAMS TO ACHIEVE GOALS				
TACTICAL EXECUTION				

Fig. 1.1 The first step in the strategic planning process is complete – a purpose has been identified and a purpose statement developed.

CHAPTER 2

Creating a Shared Vision of the Future

Vision without action is merely a dream.
Action without vision just passes the time.
Vision with action can change the world.

Joel A. Barker

In today's cutthroat marketplace, few things can be more detrimental to an organization's success than a cloudy picture of where they want to go or what they want to be. Successful organizations have a clear vision of the future supported by a focused strategy that has the agility to adapt to rapidly changing marketplaces. A clearly defined and well understood vision acts as a beacon that aligns all organizational stakeholders to a common destination.

An organization without a vision is like playing chess without understanding the end goal. Without knowing that the point is to capture the King, pieces are just being moved around on a board. A lack of vision is also like golfing without knowing that the point is to land the ball in the hole in as few shots as possible – if this is not understood, people would end up just hitting a ball around on grass leaving no point to the game. Another analogy is renovating a house

Purposely Profitable: Embedding Sustainability into the DNA of Food Processing and Other Businesses, First Edition. Brett Wills.
© 2016 John Wiley & Sons, Ltd. Published 2016 by John Wiley & Sons, Ltd.

without having a common vision of the final design. Everyone focuses on doing what they feel looks good and the house ends up looking like some fun house at a carnival. While a lack of vision or clear picture of the end destination is not common for everyday things like playing chess, golf, or renovating a house, for some reason it is very common for a large number of organizations. It is staggering to see how many organizations operate without having a clearly defined and understood picture of what they are trying to achieve.

Visioning is a basic principle that people routinely practice throughout life, most never even realizing they are doing it. For example, as kids we dream of what we want to be when we grow up. As teenagers we clarify this dream or vison and begin to take steps to bring it to reality. This may involve enrolling in a particular school or program, signing up as an apprentice, or gaining experience through part-time jobs or volunteering.

While many may not think about personal visions or dreams in the same fashion as a corporate or organizational vision, it is essentially the same thing. A future desired state. It is critical to success. On a personal level, if we never dreamed or thought about the future, we would aimlessly wander through life with no real direction and make random choices based on our feelings that day.

Sure, some folks would get lucky and stumble upon great opportunities or careers, some will fall into positive outcomes, but for the most part things do not turn out nearly as positively as they would if there was a clear vision of the future.

The interesting thing is that many organizations today struggle with this basic concept. Furthermore, where most organizations fail is not necessarily in the development of a vision or a vision statement, failure occurs by not properly leveraging the vision once it is created. Sure, crafting a clear, memorable, and inspiring vision statement is a critical piece of the puzzle and we will look at how to quickly and effectively do this below, but the real magic comes from what is done with this vision once it is hanging in a nice frame on the wall.

Companies such as 3M, Coca-Cola and General Electric are great examples of companies that have long held a clear picture of where they are headed. Relentless pursuit of the vision has helped them grow into elite organizations delivering superior long-term performance.

A vision is a rather straightforward concept. It is, quite simply, a future desired state, a picture of what success will look like at a particular time in the future, the organizational compass that points everyone in the right direction. It encompasses answers to many questions such as: What does the organization look like in the future? How big is it? What is it famous for? What size sandbox will it play in? What does it need to be?

Articulating a clear picture of the future can be challenging. Typically, there are bits and pieces of the picture but it is not clear exactly what it looks like. While it may be hard to paint an exact picture of what an organization will look like in the future, it is important to sketch out the essence of the picture.

2.1 Crafting a meaningful vision statement

Developing a vision statement has received a bad rap over the past decade or so. Perhaps this stems from pricey consultants overcharging to sequester company executives for days on end in an off-site boardroom somewhere only to develop a vision statement that acts as industrial wallpaper, collecting dust and never to be used again. Forget about this old-school approach. Developing a clear, inspirational vision statement can be done relatively quickly using the simple process that follows.

As with all steps throughout the strategic planning process, ensure the right people are in the loop but do not overcomplicate things by involving too many people.

Leveraging the same folks used in developing the purpose works well; however, remember that it is the organizational leader who should really take the lead in developing the vision. Also, make sure that someone is capturing all the ideas, work, and outcome from each step in the exercise using a whiteboard, chalkboard, or chart paper.

Step 1: Setting the stage
When developing a vision statement there is no scientific formula per se but here are some questions to help get the discussion started:
1 How big does the organization want to be relative to the rest of the industry?

2 What sandbox does the organization play in? In other words, what geographical areas will the organization operate in?

3 What will it be recognized for?

4 What services and/or products does it offer?

5 What services and/or products could it offer?

6 What does the organization want to be when it grows up?

7 What does the organization need to be when it grows up?

8 What impact will it make?

9 What benefits will it provide?

10 What are its differentiators?

The questions can go on and on but the point is that organizations must answer these questions and others to begin clarifying what the future of the organization will look like. While every organization will have a unique picture of the future, a great vision is inspiring, clear, easy to understand, motivating, and gets people fired up. Keep this in mind along with the answers to the earlier questions when moving on to the next step.

Step 2: Key word development

Using the answers from the previous step, brainstorm a list of key words that capture the essence of what the organization would ideally look like in the future. To reiterate, the key words developed here will describe the picture of what the organization will look like, or have accomplished, or be known for 10–20 years down the road. Also do not be afraid to use simple basic tools such as a thesaurus to better articulate the attributes and end goals. At this point there are no right or wrong key words; just keep the ideas flowing until there are at least 15–20 key words. Continuing with the example of Smithville Foods, this is what their list of key words could look like after completing this step:

Industry benchmark	Trusted	Leader	Disruptive
Quality	Preferred	Innovative	Cutting edge
Natural	Premium	Agile	Authentic
Produce	Fair	Quick	Reliable
Fresh	Honest	Responsive	Partner

Step 3: Key word grouping

Group the key words into similar or like categories that articulate the same idea or thought. This can be done by marking a symbol or number beside each of the similar key words as shown here:

1	Industry benchmark	2	Trusted	1	Leader	8	Disruptive
4	Quality	2	Preferred	8	Innovative	8	Cutting edge
3	Natural	4	Premium	6	Agile	2	Authentic
5	Produce	2	Fair	6	Quick	2	Reliable
3	Fresh	2	Honest	6	Responsive	7	Partner

Step 4: Key word identification

With each of the key words grouped into like categories, the next step is to choose the best word to represent each category. (These keywords will be the words used in the next step to develop a draft vision statement) For example:

1	Leader	2	Preferred	3	Fresh	4	Premium
5	Produce	6	Agile	7	Partner	8	Innovative

Step 5: Drafting a vision statement

With a distilled list of key words, the next step is to use these key words to build a one-sentence draft vision statement. To do this, string together each of the key words using various verbs or action words such as "through", "by" "deliver", "provide," and so on. Try developing 3–5 different statements. Try avoiding things like "the largest" or "the best" in the statement. Instead, use descriptors that will lead to this. For example, if one is the most trusted partner or provider, this would lead to growth and potentially being the largest or best. Using the key words from the previous step, the following could be some potential vision statements for Smithville Foods:

1 **To be the North America's preferred partner for fresh and innovative produce solutions.**
2 **To be North America's premium provider of fresh produce.**
3 **North America's most preferred produce provider.**
4 **The greatest food company in the world.**
5 **To provide innovative and fresh produce solutions.**

Step 6: Finalizing the vision statement

It is advisable to sleep on or think about each of the statements for a few days and then come back and review them. It may also be prudent to garner feedback from key stakeholders. However, remember it is the responsibility of the leader and leadership team to define the vision so try not to overcomplicate this process by involving too many individuals. After a few days, the group should come back together, review each statement and choose the one that best articulates the spirit of what the company wants to be in 10–20 years. Be careful though … when crafting a vision statement, do not fall into the typical traps that many organizations do. For example, many vision statements are cheesy, idealistic statements that are anything but inspirational. An organization's vision statement doesn't need to state that "Company ABC will provide safe, quality food with great customer service." Versus what? "Providing toxic, crappy food with inferior service." Such statements show that management lacks imagination, and perhaps in some cases, direction. Continuing with the example from above, a final vision statement for Smithville Foods could look like this:

North America's preferred provider of fresh produce solutions.

Once the organization has developed their vision of the future, it must become everyone else's vision of the future. Let's look at how to do this.

2.2 Creating a shared vision

Developing a vision statement is one thing, creating a shared vision is another. A shared vision happens when all employees in the organization clearly understand the vision, buy into it, and understand their role in bringing it to life. When all the players have the same endgame in mind, beautiful things happen. Every activity, every move, and every decision made across the organization is aligned to one single end point. All the cogs in the wheel are turning in the same direction. This is the tricky part. However, following a few simple steps can turn a meaningless statement into a powerful driver of performance.

2.2.1 Painting a clear picture of the future state

To ensure that everyone understands the vision, it is the responsibility of the organization's leader to not only clearly communicate the vision but also paint a vivid picture of what the organization will look like once the vision has been achieved. Let's look at how this can be done using Smithville's vision statement:

North America's preferred provider of fresh produce solutions.

In order to clearly communicate and paint a picture of what the organization will look like when this vision is achieved, leaders must first understand themselves, what this means. Asking some pointed questions can help to clarify this:

What does it mean to be preferred?

It could mean cheapest prices, quick delivery, superior quality, innovative products, etc. The leader(s) must decide what it means to the organization, but for this example let's say it means to be all of these things. So in order to be North America's preferred partner of fresh produce solutions, Smithville Foods will continually offer new, exciting produce solutions that are brought to market in record time from growers at a price that equals or betters competitors.

What is meant by provider?

It could mean grower, processor, distributor, etc. Again, leadership must identify what this means. In this case let's say a provider means a processor and not a grower or distributor. By providing this clarification, it provides focus and sets boundaries for how the organization will develop and grow.

What is meant by produce solutions?

This could mean traditional fresh produce solutions such as tomatoes, peppers, lettuce, etc., but will grow to offer other fresh produce as time evolves.

Once leadership is clear on what the vision means, and what that picture really looks like, it must be clearly communicated to stakeholders.

2.2.2 Aligning daily activities to the vision

The power of a vision really starts to manifest when all the little daily activities, decisions, actions, etc. align with and support the overall vision. However, in order for this to happen, people need to understand how their daily, what may seem like mundane tasks contribute to achieving the overall vision. Continuing with this example, let's look at a few roles and see how they contribute to the overall vision.

For example, let's look at a frontline worker at Smithville who packs the produce into bags before being shipped out to the customer. The more efficiently this person can safely package the product, the more successful the organization will be in bringing the product to market in record time, which will help Smithville become "preferred." Also, as the last point of contact before the customer, they have the last chance to ensure that the product is fresh – also a part of the vision.

Let's look at another role. Say the marketing folks. Well, they need to ensure that all the brand messaging aligns with the vision. It must portray all the characteristics and qualities noted in the vision.

Buyers need to understand their role as well. Since "preferred" also includes pricing, then buyers play a critical role in securing the lowest prices and best contracts; otherwise, they will not be "preferred" and therefore Smithville will fail to achieve their vision.

Every person in the organization must understand and be able to directly link their daily activities to achieving and supporting the vision.

The only way to really ensure that all folks across the organization understand the vision and their role in achieving it is to take the time to talk with them. Sure, in larger organizations it is difficult for one or a few leaders to talk to every employee but this task can be cascaded down to directors, managers, and so on. Having said that, senior leaders would greatly benefit from taking the time to do it themselves.

Creating a shared vision takes some practice but over time, becomes easier and more natural. The key to success here is constant reinforcement. This is not something that can be done once or twice, it needs to be done over and over and over again, from different angles and through different examples. There needs to be constant reminders along with recognition or repercussions if actions do not support the vision.

The world's most successful leaders are masters in developing a shared vision and motivating their employees to support it. Richard Branson, Steve Jobs, and Donald Trump were/are all masters at this. No one at Virgin would ever think about offering a product or service the same way a competitor does, no one at Apple would design a complicated, clunky device, and no one in the Trump organization would ever dream about building a second rate hotel. This is because everyone who works at these organizations understands what the organization is shooting for, what it stands for, and how they contribute to bringing the vision to life.

Now, it is important that there is a clear link between achieving the vision and fulfilling the purpose. The organization must ask itself if achieving the vision would support fulfilling the purpose. In the case of Smithville Foods, the link is clear. If they are the preferred provider, they will maximize the amount of people who consume their products and if their products are fresh produce solutions, it supports health and well-being. Thus their vision of being the preferred provider of fresh produce solutions clearly supports their purpose of improving the health and well-being of current and future generations.

At this point in the overall strategic planning process, the organization has a clear picture of the future linked to a meaningful purpose. This is powerful as it will guide, support, motivate, and inspire employees to align daily decisions and activities to a common endgame. All the cogs in the wheel start moving in the same direction, all efforts are aligned in a common direction leading to enhanced performance and superior results.

Progress indicator

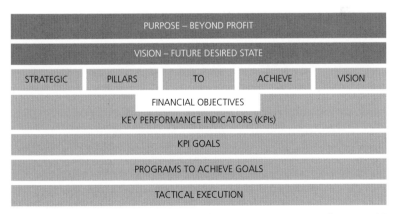

Fig. 2.1 The second step in the strategic planning process is complete – a vision has been developed and articulated in the form of a vision statement.

CHAPTER 3

Getting Focused – Pillar Development

My success, part of it certainly, is that I have focused in on a few things.

Bill Gates

Historically, organizations have been guided towards their vision through a mission, usually articulated in the form of a mission statement. This mission sets out the key focus areas or the "guiding lights" that will steer the organization closer to achieving its longer term vision. Unfortunately, the traditional mission and corresponding mission statement are too often narrow in focus, static, and fail to incorporate all three dimensions of Sustainability – leaving a lot of value on the table and ignoring stakeholder expectations. Often these missions focus on things like customer service, quality, price, and so on – all important but not very inspiring; in today's world, these types of things are expected.

Fast forward to the late 1990s and early 2000s when organizations who understood the importance of Sustainability begin developing Sustainability focused "mission statements" as part of an overall "Sustainability Strategy". These statements, like their counterparts mentioned earlier, tend to be narrowly focused, in this case, on certain aspects of Sustainability such as the environment or a social issue. Great intention here and one must start somewhere but this approach

Purposely Profitable: Embedding Sustainability into the DNA of Food Processing and Other Businesses, First Edition. Brett Wills.

can actually have a negative impact by siloeing Sustainability as an initiative or project that is outside the core focus of the business.

Success in today's marketplace requires a focus not just on the single bottom line of profit but on the triple bottom line where social, environmental, and economic performance work together to maximize organizational performance and in turn, maximize profit. In order to do this, a fresh approach is required, one where there is not only a clear focus for the organization but a focus that addresses all dimensions of Sustainability by seamlessly integrating each dimension into the overall strategy.

An elegant approach for doing this is through the use of strategic "Pillars". This concept of strategic pillars is not necessarily new; there are many strategic planning models that use the approach of pillars to concentrate focus such as the Rockefeller Habits. These strategic pillars support performance towards the vision to ultimately fulfill the organization's purpose.

3.1 The power of pillars

For centuries, builders and architects have used pillars to provide support and structural integrity to a building. Without these pillars, the building becomes unstable and eventually collapses. While there is much more that goes into constructing a stable building, without pillars to support the additional elements, the building collapses. Recognizing the role a pillar plays in a sound structure, societies across the world now use them as symbols of strength and unity to convey a message of power or to bring people together.

When it comes to business, the use of a strategic pillar serves the same function in building a stable business as a structural pillar does in building a stable building. Adopting the concept of strategic pillars lends tremendous value to driving performance in a number of ways. Let's look at few of the main ways the use of strategic pillars can power organizational performance.

3.1.1 Providing clear direction and focus

One of the main benefits that comes from the use of strategic pillars is in providing clear focus and direction for the organization. In today's highly competitive marketplace, organizations are continually being pressured to find innovative solutions to grow the top line, reduce costs and drive profit.

In response to these pressures, more and more organizations are losing focus on what they do well by choosing to go outside their core competency areas to find what are perceived as "better" ways to drive profitability. The result is short-term gains at the sacrifice of long-term success as they try to be more things to more people. There are many examples across a wide spectrum of industries that illustrate how organizations are deviating from their core competencies. It can range anywhere from entering brand new industries of which they have no experience or entering brand new geographical markets (think Target's failed move into Canada) to offering products instead of services to leaning on technology instead of their people.

When organizations lose focus on what they do well, it can have a number of negative implications to overall performance. For starters, it creates distraction within the organization. Resources are diverted in an attempt to foster new initiatives and as a result, well-performing departments, product lines, people, etc., suffer from the distraction. While results may not immediately show downward performance, over time a lack of attention to these once, well-performing areas, begins to show itself in the form of downward performance. While some companies can recover, often the damage has been done and there is no or very limited recovery.

Beyond distraction, losing focus can also have other negative repercussions, particularly to the brand. Decreased brand awareness results as people struggle with identifying what the organization is good at. Loss of brand equity happens as people view the organization as a jack of all trades and master of none.

Beyond brand, there are additional negative repercussions such as the inability to attract top talent as folks look to build careers with industry leading organizations, not ones with an identity crisis. All of these factors result in the eventual loss of market share, shrinking margins, and overall loss of profitability.

Now, it is important to note the difference between a loss of focus and expanding or adapting focus to better serve customers and meet stakeholder expectations. Adapting or expanding the business is great but must be done in a way that is natural, organic, and supports rather than deteriorates the brand.

3.1.2 Creating a culture of Sustainability

In Sustainability circles across the globe, there is much talk about "embedding Sustainability into the DNA of the organization" – for many Sustainability professionals and executives passionate about Sustainability, this is the so called "holy grail". Beyond Sustainability, many organizations are also focused on embedding other changes into organizational culture. The annoying thing is that many folks readily spew this off their tongue but have no idea what it actually means or how to do it.

So what does this really mean to embed Sustainability into the DNA of the organization – aka embedding Sustainability into organizational culture? At the end of the day what this really means is that Sustainability becomes part of daily decision making. Whether it be minor decisions, major decisions, strategic decisions, or otherwise. In other words, every decision made now includes social, environmental, and economic considerations.

Take a simple example; procuring a new piece of equipment. Many organizations routinely consider things such as quality, productivity, return on investment (ROI), and even things like safety are commonly part of a procurement decision. If Sustainability were embedded into the DNA of the organization, the decision on what piece of equipment to buy would include all the these things but now also include Sustainability-related criteria such as energy consumption, types of materials used to make the equipment, where it is manufactured, how it is manufactured, and so on.

Another simple but powerful example relates to behavioral decisions. People make hundreds of behavioral decisions on a daily basis, often without even recognizing that a decision has been made. Shutting off a light when leaving a room, for example, or shutting off a machine during lunch or a break are behaviors that require a decision – to do it or not to do it. Often, the decision has been made so frequently that it is no longer a conscious decision and becomes automatic. If Sustainability were part of the "DNA," i.e. part of daily decision making, people would automatically make the decision to turn off the light off or switch the machine off during lunch. Obviously, very simple examples are given here but the concept should be clear; embedding Sustainability into the DNA requires that all decisions improve social, environmental, and economic performance instead of just financial performance.

The challenge is how to actually get folks to make Sustainability part of daily decision making. This is where the pillars come into play; they are the first step towards embedding Sustainability into organizational culture or embedding anything into organizational culture – it is not the be all and end all, but a critical first piece to the puzzle (other pieces of the puzzle are addressed throughout the remaining pages and chapters). The use of strategic pillars not only sets direction and focus for the organization it also enables the three dimensions of Sustainability to be integrated into the overall strategy. Let's take a look at how this is done.

3.2 Building the pillars

Let's take a look at how to develop a set of strategic pillars that will keep the organization focused and successfully integrate all three dimensions of Sustainability – this is where the magic happens. The first two steps in this process are unique but the remaining steps are similar to those used to develop the vision. Before getting started, it is crucial to understand that this **is not about building pillars of Sustainability**, **it is about building pillars that incorporate Sustainability.**

Step 1: Pillar identification

The first step in building a set of strategic pillars is arguably the most challenging. Identifying the right pillars of focus is difficult for a few reasons. First off there is no scientific approach to identify the correct pillars. Secondly, there is no immediate feedback on success; only performance over time will tell if the pillars were correctly identified or not. The great thing is that this Strategic Planning System is flexible enough to allow for mid-course corrections to the pillars by working back through this step in the process to adjust as needed. Maybe the adjustment is driven by lack of performance or maybe it is driven by changes in the marketplace. Regardless of the reason, the pillars can be modified to change the path while still ending up at the same destination. In other words, the end goal is the same but the path to achieving it changes. Now, this is not about flip-flopping and constantly changing course because immediate results are not favorable; organizations are encouraged to stay the course but if required, the focus or path can be modified through pillar modification.

While there is no scientific approach to correctly identify pillars, here is some guidance that will help to correctly identify the right pillars:

The pillars should:

- Be broad in focus but clear in direction
- Typically consist of 3–5 pillars – any more, focus can become diluted and any less, focus can be too broad
- Be specific to the organization, not to the industry, competitors, etc.
- Be aspirational but at the same time realistic
- Support the vision

With these guidelines in mind, the next action is critical – identifying the actual pillars. To identify the pillars, there is essentially one question that needs to be answered:

What are the key areas the organization must excel in to achieve the vision?

Answering this question can be tough; here are a few supporting questions to get the ball rolling and spur discussion to help answer the larger question:

- What are the organization's differentiators?
- What does the organization do well?
- What has driven success historically? Will this continue to be the case?
- What are major trends in the industry?
- What are stakeholder expectations?
- What would disrupt the industry?

Remember that the pillars are unique to the organization and the types of pillars will vary from industry to industry. For example, a business-to-business manufacturing company may identify pillars that are more focused on operational excellence, supply chain, or products offered, maybe even innovation or research and development. A hospitality company may identify pillars that are more focused on customer experience, brand, and services offered, so on and so forth.

So really try and focus on the core competencies of the organization, the things that have made the organization successful, *and* areas that may not have been a focus in the past but are critical to success going forward.

With all this in mind, it is time to actually begin identifying the areas of focus that will then be distilled down to a set of pillars. A great way to work through this is to simply gather the key decision makers in the organization together to brainstorm the focus areas. At this time, do not worry about fancy names for the pillars – this comes later – just focus on identifying the areas of focus.

Using the Smithville Foods example, let's work through this step to identify some potential focus areas for this organization.

By answering the earlier questions and keeping the vision and the purpose in mind, here are some potential focus areas:

Employees	*Product innovation*	*Competitive pricing*
Stable supply	*Community support*	*Environmental performance*
Food safety	*Brand awareness*	*Customer loyalty*
Quality	*Operational excellence*	*Healthy products*

Now, look at the different focus areas. Do any of these overlap or could they fall under one larger focus area? For example, food safety, quality, and environmental performance may all relate to operations. So these three focus areas could fall under one pillar of "operational excellence."

Competitive pricing, product innovation, and healthy products really all relate back to products so these focus areas could fall under one pillar of "products."

Customer loyalty and brand awareness both relate back to the customer or consumer so may fall under one pillar of "customer". Some focus areas may not overlap and therefore require a pillar of their own such as Community support and Employees in the above example.

With the objective to find common focus areas, go back through each of the areas identified above and group into like areas or categories to come up with a set of 3–5 pillars that cover all the areas identified. Continuing with the above example, Smithville Foods may identify the following pillars based on the multiple focus areas that were originally identified:

Customers
Products
Operational excellence
People
Community

It is important to note that while every organization will have unique pillars, there tends to be similar types of pillars across similar industries. For example, manufacturers will typically all have some sort of pillar focused on operational excellence. For a service company, this may be important but not as critical and therefore would not call out a specific pillar of focus in this area. Furthermore, some pillars are consistent across all industries such as a People pillar.

Most organizations will also have a Partner pillar but some may call out specific partners that are more important such as "Customers", where other organizations may choose to just have a generic partner pillar that looks at customers, suppliers, and other key stakeholders. Remember this is more art than science and success here depends on understanding the basic fundamental concept that the pillars are the focus areas critical to success.

A final note here—at this point there is no need to worry about defining *exactly* what this pillar is focusing on. Later steps will focus on developing a "mission" statement for each pillar that will clearly articulate the focus for the pillar. At this point, it is only important to understand the spirit of what the pillar will focus on.

Note that many organizations already have strategic pillars or strategic areas of focus, this is great. Most of the hard work has been completed for this particular step. It is suggested to review the current pillars and ensure they align with the vision and are still relevant. However, the remaining steps still need to be completed in order to ensure the dimensions of Sustainability are fully integrated and the pillars are clearly defined.

Step 2: Integrating Sustainability into the pillars

Now it is time to review the pillars with a Sustainability lens. This is critical to ensuring all aspects of Sustainability have been covered across the pillars. Before getting started, there a few things to keep in mind when working through this exercise:

- The pillars together must address each dimension of Sustainability: Social, Environmental, and Economic
- Remember that Social Sustainability can be broken into three subcategories of People, Partner, and Community
- Remember that Economic Sustainability can be broken down into three subcategories of Agility, Resilience, and Innovation

(Refer to the Sustainability primer at the beginning of this book for a better understanding of Sustainability and each of the dimensions.)

With these thoughts in mind, it is a relatively straightforward process to integrate Sustainability into the pillars. It involves first understanding each dimension of Sustainability and how this translates to the particular organization and then comparing this to the pillars to ensure each dimension has been covered across the pillars. It may be covered in different ways. It could be covered because there is a specific pillar that calls out a specific dimension of Sustainability, for example an "environment" pillar. It may also be covered because an identified pillar will incorporate the dimension such as including environment in say, an operational pillar. Remember that a focus on all dimensions of Sustainability is critical to success in today's marketplace so while each individual aspect may not be super critical to success and therefore does not have a specific pillar called out as a focus area, each dimension should be covered within the pillars.

Ok, let's look at how to integrate each dimension of Sustainability into the pillars by addressing each dimension in turn using Smithville Foods as an example.

Social sustainability

As noted earlier, Social Sustainability can seem at first glance to be a slushy, altruistic concept to wrap one's head around. However, breaking it down into the three subcategories of people, partners, and community makes it much easier to address. (See Sustainability primer at the beginning of the book for more detail on the categories of Social Sustainability.)

Looking at the *People* aspect of Social Sustainability, the organization must decide if its people are critical to success. Having said that, it's hard to imagine any organization where the people are not critical to success. Assuming the organization determines that its people are critical to success, it must then review the pillars to see if any of the pillars directly address the safety, health, well-being, development, happiness, engagement, etc., of employees. Perhaps there is a specific employee or people pillar such as in the above example of Smithville Foods, in which case, all is good. However, if people or employees have not been specifically called out as a focus area then it needs to be decided if it will fit into an existing pillar or if a new one needs to be created. For example, if an operational pillar has been identified in step 1, it may make sense to incorporate it into the operational pillar. If the decision is to create a new pillar, then revise the list of pillars accordingly. If the decision is to incorporate into an existing pillar, note which pillar it will be incorporated into.

The *Partner* aspect of Social Sustainability can be a little trickier. As noted in the primer on Sustainability, partners look at suppliers, customers, and other key stakeholders. So the key here is to decide which partners are critical to achieving the vision. For some organizations, all three categories of partners may be critical to success and in this case it may make sense to call out a specific pillar for Partners. For others, certain partners may be more critical than others. A service company for example may recognize the importance of suppliers but will not necessarily make or break the organization where a focus on customers is a critical component to success and achieving the overall vision. In this case, there may be a specific pillar for customers. Once it has been decided which partner(s) are critical to success, review the pillars and if there is no pillar calling out the specific partner(s), then again, one may need to be created or alternatively, the organization must identify which of the existing pillar(s) the partners can fit into. For example, if it is decided that both suppliers and customers are critical to success, perhaps suppliers could fit into an operational pillar while a separate pillar may be created just to focus on customers. Again, note what the decision is.

Community is the final piece to look at under the lens of Social Sustainability. In today's marketplace, it is more important than ever to have a focus on giving back and strengthening the community or supporting a social cause as discussed throughout this book. Once again, the organization needs to determine if this is a critical component to success. It is strongly suggested that organizations include a focus on community for the many reasons presented in this book. Assuming that community is a critical piece to success, review the pillars to see if it has been covered anywhere. If not, decide if it requires its own pillar or if it can be incorporated into another pillar. For example, some companies will choose to incorporate giving back or supporting a social cause through their products or services and some through their employees volunteering, so in these cases it would be integrated into those pillars. Whatever the decision, take the necessary action by revising the pillars or noting which pillar(s) it will be incorporated into.

Environmental Sustainability

The environmental dimension of Sustainability is a little easier to integrate into the pillars than Social Sustainability. Like the Social Sustainability examples discussed earlier, the organization must first decide if a focus on the environment is critical to success. Remember, we are not talking about hugging trees here, it is about cutting costs and driving revenue through things like using less energy and water, sending less waste to landfill, designing products for re-use, developing innovative new products, and so on. In today's marketplace, every organization needs to seriously consider environmental performance as a key ingredient to success.

Once the organization identifies environmental performance as a critical factor for success, the pillars need to be reviewed for inclusion of the environmental dimension. The rub here is that the environment is a broad concept that may refer to many things such as carbon, water, waste. etc., and many areas such as operations, products, supply chain, and so on, making it difficult to integrate into the pillars. To overcome this challenge, organizations should at this point, have a good idea of where their biggest environmental impacts stem from, i.e operations, products, services, supply chain, etc. However, at this point, try not to get too caught up in the details of what the specific environmental impacts are; there is time for that later in the process. At this point, simply focus on the larger picture of where the environmental impacts are coming from.

With an understanding of where the major environmental impacts lay within the organization, review the pillars to determine the best way to integrate the environment. Like the social side, it may require a pillar of its own or it may be integrated into one or more of the existing pillars. Many times, organizations will call out a specific environmental pillar but just as popular is integrating into another pillar. A general guideline here is that if the organization's environmental focus is more around internal operations (i.e. energy conservation or waste diversion) then it can be integrated into an operational type pillar. If the focus is more around reducing the environmental impacts of a product, service, or something like packaging, the environmental piece can be implemented into a product or service-related pillar. In the many cases where the greatest impact comes from the supply chain, it may make the most sense to integrate environment into the pillar that focuses on the supply chain.

For organizations that are focused on reducing environmental impacts across the entire value chain, it may make sense to call out a separate pillar for the environment. Alternatively, the organization can embed environment into all the pillars where there are material environmental impacts. There are many possibilities here and the ideal solution will vary from organization to organization. For example, if a distributor or marketer of a certain product outsources manufacturing and therefore does not have an operational pillar but maybe a partner or even specific supplier and customer pillars along with a product pillar, environment may be integrated into each of the pillars.

Remember, the spirit of what is being done here. The point is to first determine where the greatest environmental impacts stem from and then ensure that the pillars which focus on those areas have environment built into them. If a source of environmental impact is not addressed through a pillar, this may spur the development of a pillar for that area or to have that area addressed within an existing pillar.

Economic Sustainability

As discussed in the Sustainability primer Economic Sustainability is not what it appears to be. The name suggests it is about finance, this is not the case. Economic Sustainability is about what an organization is focusing on today to keep the doors open 5, 10, 20 even 50 years down the road. It is about what the organization is working on today to have healthy financials tomorrow.

Sure, this seems very "blue sky" and complex but the concept is sound and as discussed in the Sustainability Primer, the notion of "Economic Sustainability" will drive long term success through three high-level concepts; Innovation, Agility and Resilience.

Let's break this down into bite-sized chunks by looking at each of the three concepts in turn to see how Economic Sustainability can be integrated into the pillars.

The first of the concepts we will look at is innovation. This is a big buzz word in business these days. Organizations across the globe are increasingly talking about how they are innovating products, solutions, processes, and so on. However, very few organizations are successful innovators and many lose focus after a relatively short period of time, while others can be very successful in this area, like Apple or Dyson. This begs the question – What is the difference between those who are successful innovators and those who are not? The high-level answer is quite simple actually; the successful organizations have embedded innovation into their "DNA" or culture like other aspects of Sustainability and therefore it becomes a focus. Once a focus, the structure and support mechanisms are established to drive performance in this area.

As discussed earlier, the pillars are the first step for embedding a concept, view, or belief into the DNA, so ensuring innovation is covered across the pillars will be the foundation for embedding this behavior into organizational culture – the structure and support mechanisms for driving innovation (and the other pillar focus areas) will come in later steps that are covered in Chapters 4–7.

The process here is the same as the other aspects of Sustainability, first decide if innovation is critical to the success of the organization. It should be – organizations who are not innovating are dying. Now, for some organizations it may be more critical than others and therefore it may deserve its own pillar. For other organizations, it may not necessarily deserve its own pillar but is important enough to be integrated into an existing pillar. For example, a business-to-business manufacturer that is not consumer facing and simply building to customer specification, may not have innovation as a separate pillar. Rather the manufacturer may include it within an operational pillar to drive focus on developing innovative processes to better build customer products vs actually designing innovative products. On the other hand, the organization that actually designs the product and outsources manufacturing, may indeed have a separate pillar for innovation – they may also just include it under a pillar focused on say "products" or maybe a pillar focused on "design" if they feel it does not require its own pillar. The organization must decide where innovation will bring the greatest value and integrate into that pillar or call it out as a separate pillar because it is just that important. Whatever the organization decides, it must be noted by either revising the pillars to call out innovation separately or noting which of the existing pillars will need to have innovation integrated into them.

Agility is the second concept to look at under Economic Sustainability. Agility is the ability to quickly react to changes in the marketplace whether it be consumer preferences, technology, changing demographics, or otherwise. Like innovation, some organizations are good at reacting to changes in the marketplace like PepsiCo who recognized the trend towards more health conscious eating. Others are unable to react to changes for one reason or another such as the brick and mortar video stores that did not react fast enough to the online streaming of movies.

Before getting started with assessing the importance of agility to the organization, it is important to remember that the heart of agility is having adaptable processes, equipment (capital), technology, and people, so keep this in mind when working through this exercise. Sure, a lot of times it is the product or service that needs to adapt, but it is the previously mentioned things that actually produce the products and services. Once again, the organization must decide if agility is important and in what areas it applies to. Once this is has been decided, return back to the pillars to see if/where it has been addressed. For some organizations such as design or technology firms, they may see agility as so crucial to success they actually call it out as a separate pillar. Others may recognize the importance of agility but maybe it is more relevant in certain areas and therefore may choose to integrate agility into an existing pillar. For example, a food company may choose to integrate it into a product pillar to drive the ability to react to changing trends in food preferences. The business-to-business manufacturer from earlier that is building to customer specifications may choose to integrate agility into the operational pillar in order to react to design changes coming from customers. Regardless of the decision, add a pillar if creating a new one or note which of the existing pillars require the integration of agility.

The final piece of the Economic Sustainability puzzle is *resilience* which is the ability to recover after a business disruption. The disruption may be a collapse of the stock market, maybe its falling oil prices, a natural disaster, or the sudden loss of the organization's founder. Regardless of the type of disruption, organizations must have the ability to recover and this is done through a focus on decentralization – decentralization of decision making, physical or geographical decentralization, decentralization of investments, decentralization of knowledge, information, systems, and so on.

Every organization is subject to disruption, some more than others, but every business is subject to some type of disruption. Like with the other concepts of Sustainability, organizations must first decide if this is something important enough to focus on. If so, the organizations must first understand where/what/when the potential disruptions could be. Understanding this, organizations must then decide if it is important enough to call out as a separate pillar or just requires integration into an existing pillar. For example, an organization that has a rigid top down hierarchy where everything is kept within the executive suite may see decentralization and therefore resilience as a critical component that must be applied across every facet of the organization in order to change the culture and therefore call out resilience as a separate pillar. (They should also make sure the executive team does not travel together.) Or, an organization such as Apple or even Dyson, where arguably the main thrust of success comes from the founders, may choose to integrate resilience into the area or pillar they influenced most such as a product pillar or design pillar. For the final time, once the organization decides on whether to call out a separate pillar or integrate into an existing one, note the decision.

At this point, the organization should have a very clear idea of where each aspect of Sustainability is addressed, either as a separate pillar or integrated into one of the existing pillars. Let's take a look at what the end product may look like for Smithville Foods.

Customers (partner aspect covered here)
Products – innovation, environment, and social cause (community)
Operational excellence – innovation, environment, and suppliers
People – innovation and agility
Community – scratch this, added to the products pillar
New Pillar – Resilience

Step 3: Pillar key word development

With the pillars identified and a clear picture of how each dimension of Sustainability will be addressed across the pillars, it is time to clearly articulate the focus for each pillar by developing a mission statement for each pillar. The exercise for doing this is similar to developing a vision statement and starts by developing a list of key words that capture the essence of what the pillar is focused on and aspiring to achieve, while including the aspects of sustainability that fit into that pillar. It is important to keep in mind the vision and purpose when working through this exercise to ensure the pillars support and align to them.

Going back to the whiteboard again, begin brainstorming a list of key words that represent what the pillar is ultimately trying to achieve. Using Smithville Foods as an example, let's work through this step.

Pillar: Customers

Loyal, customer service, agile, responsive, fair pricing, transparent, nimble, relevant, partnership, mutually beneficial, foster, on time, delivery,

With an understanding of what is important to their customers and thinking about what the customer would require to choose them as their preferred supplier, the above key words were developed to capture the essence of what this pillar is focused on and aspiring to achieve.

Pillar: Products

Innovation, environmentally positive, social benefit, healthy, quality, safe, cost effective, convenient

In developing the key words for this pillar, Smithville Foods looked at what people love about their products, what attributes they and others would like to eventually see and how their products will help support the overall vision. Note that the aspects of Sustainability identified in the last step to fit under this pillar are included in the list of key words. This is an important piece to remember in this exercise. If the aspects of Sustainability are not built into the list of key words, they will not make it into the mission statement and therefore be lost as a focus. If they are lost as a focus in the pillar, the whole point of leveraging the pillars as the foundation for embedding Sustainability will also be lost.

Pillar: Operational Excellence

Quality, safe, productivity, continuous improvement, environment, innovation, stable supply, cost effective, automation, efficiency

When looking at an operationally focused pillar, organizations should be looking at how the operations can support the other pillars along with how it specifically contributes to the overall vision while including the aspects of Sustainability that have been identified to fit under this pillar. This may seem like a lot but it can usually be captured with a small group of key words like the Smithville Foods example earlier. In this pillar and in others, some things may seem obvious and therefore there is a tendency to exclude them from the list of key words. Remember though, these pillars are the key focus areas and they contain the key elements for success so excluding what can seem obvious or basic may in fact cause the organization to lose focus on those things and cause problems down the line. Point being, if something is important to the area the pillar covers, even if it seems obvious or basic, write it down, it can always be excluded from the final statement if need be.

Pillar: People

Engaged, safe, happy, motivated, passionate, accountable, dependable, innovative, agile, trustworthy, loyal, foster, develop

---

The people pillar is a common pillar amongst successful organizations and generally tends to focus on similar things such as engagement, accountability, safety, etc. Often where these pillars differ amongst organizations is in which aspects of Sustainability they address. In the case of Smithville Foods, they felt that innovation and agility where two aspects of Sustainability that were particularly relevant for their people and thus have included those as key words here.

Pillar: Resilience

Recover, respond, risk mitigation, adapt, change, processes, systems, de-centralized

---

During the last step, Smithville Foods decided that resilience was important to their organization and particularly important to fulfilling their overall purpose. They must be around for a long time in order to impact the well-being of current and future generations. For this reason, they added resilience as a pillar. In their case, they feel that there is not one particular area where this has more impact so the key words capture the essence of what this pillar is trying to achieve across the entire organization.

By this point, the organization should have a pretty good idea of what each pillar is all about and what it is trying to achieve and aspire towards. Like with each of the exercises for each step in the overall process, make sure these key words are captured somehow, either by having someone transfer them to a document or by taking a picture of the whiteboard, flip chart, etc. Having all this information is helpful should the organization want to go back and change something.

Step 4: Key word grouping

With the key words laid out for each pillar, the next step is to distill down the key words into like or similar categories that share the same intent or spirit and then choose the best word to articulate the category – similar to what was done with the vision statement. The same numbering process that was used for grouping the words into like categories to build the vision statement can be used to group the key words together for each of the pillars – all the words to be grouped together receive the same number and therefore are in the same category. Again, like with the vision statement, some words may not fit into a category and therefore will stand on their own. Remember the intent here is to distill down a large list of key words to a smaller list that can be incorporated into a mission statement for the pillar. To avoid redundancy, we will skip showing what the grouping of the words looks like but the final result should look something like the Smithville Foods example here.

Pillar: Customers

Loyal, nimble, customer service, transparent, partnerships

Pillar: Products

Innovation, environment, social benefit, quality, cost effective

---

Pillar: Operational excellence

Quality, continuous improvement, environment, innovation, stable supply, efficiency

---

Pillar: People

Engaged, safe, motivated, dependable, agile, foster

Pillar: Resilience

Respond, risk mitigation, systems, de-centralized

After grouping the key words for each of the pillars into like categories and choosing the word that best articulated the category, Smithville Foods was left with the key words listed for each pillar. Organizations working through this exercise should have something similar to the Smithville Foods example.

Step 5: Developing pillar mission statements

With the key words developed for each pillar, it is time to work these words into a draft statement. This is the opportunity to clearly articulate the focus and direction for the pillar. It can be thought of as an aspirational statement of what the pillar is ultimately trying to achieve. Use the same process that was used for drafting the vision statement, string together the key words to form 3–4 draft mission statements for each pillar. Then choose the best statement or cut and paste different pieces of the statements together until one final statement can be decided on. Continuing with the Smithville Foods example, here are some examples of what the pillar mission statements could look like using this process:

Pillar: Customers
Keywords: Loyal, nimble, customer service, transparent, partnerships
"Build loyal partnerships with our customers through a focus on radical transparency and nimble customer service."

Pillar: Products
Keywords: Innovation, environment, social benefit, quality, cost effective
"Offer quality, innovative products that are soft on the environment and provide social benefit all at a fair price."

Pillar: Operational Excellence
Keywords: Quality, continuous improvement, environment, innovation, stable supply, efficiency
"Efficiently provide a stable supply of quality food products through a focus on innovative processes, continuous improvement and environmental sustainability."

Pillar: People
Keywords: Engaged, safe, motivated, dependable, agile, foster
"Foster a safe workplace full of engaged and agile employees who are motivated to achieve our vision."

Pillar: Resilience
Keywords: Respond, risk mitigation, systems, de-centralized
"Develop decentralized systems that mitigate risk by quickly responding to business interruptions."

It may be difficult to work some of the chosen key words into a cohesive statement that flows well. If this happens, simply go back to the original list of key words and try substituting another word from the same category to see if that works better.

Even after the statements have been drafted and seem to flow well, they may need some word-smithing and fine tuning before they are finalized. It is suggested to 'sleep" on the draft statements for a few days and come back to them before finalizing.

At this point, the organization may choose to develop a more creative name for the pillars. This can be done in many ways and may be an exercise for the marketing department. Regardless of who completes the exercise, organizations should try and tie the pillar names back to the company's overall branding and/or industry. Typically the same word(s) are used for each pillar and should be a verb. For example a food company may want to incorporate a "food" word like grow or nourish into the pillar name. For the example of Smithville Foods, this could look something like the following:

Nourishing our customers
Nourishing our products
Nourishing our operations
Nourishing our people
Nourishing resilience

A manufacturing company may want to incorporate words relevant to that industry such as building or delivering or producing.

Organizations may also choose to develop a logo or icon for each pillar that visually illustrates the spirit of that pillar.

A creative name and logo or icon for each pillar does add some good value. It makes the pillar instantly recognizable, easier to remember, and more engaging for both internal and external stakeholders.

3.3 Visually illustrating the pillars

With the hard work of building the pillars completed, now is a good time to design a way to visually illustrate the pillars in a picture or diagram. Visually illustrating the pillars can be useful for a number of reasons. It allows quick communication of the pillars, can be used in presentations, brochures and other marketing material, better engage visual learners, and so on. There are infinite ways to visually illustrate the pillars but here are a few standard approaches.

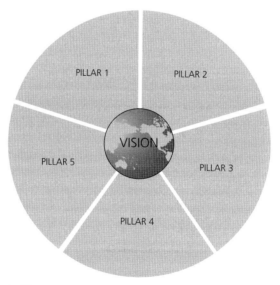

Fig. 3.1 Wheel pillar.

The Wheel
This is an increasingly popular way to illustrate pillars as the wheel denotes continuous improvement. With the vision in the center, it visually communicates that the pillars support the vision.

The House
This diagram has been used for decades in the continuous improvement world and is still relevant today. It clearly indicates that the

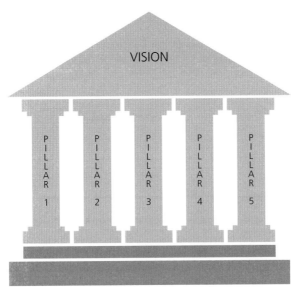

Fig. 3.2 House pillar.

pillars support the vision but also has space to include core values or other information.

The Puzzle
This format shows how all the pillars fit together to complete the puzzle and achieve the vision.

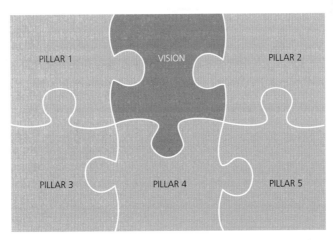

Fig. 3.3 Puzzle pillar.

At this point, the organization has a very solid foundation on which to develop the remainder of the strategic plan. It has clearly identified the areas of focus that are critical to organizational success in the form of pillars and pillar mission statements. At the same time, all dimensions of Sustainability have been successfully embedded into the strategy and now have the beginnings of a strategy that is sustainable.

The remaining steps will use these pillars to build out the remainder of the strategic plan starting with the development of financial objectives and key performance indicators that will be used to measure financial results and performance under each pillar that will gauge progress towards achieving the aspiration set out in each pillar mission statement.

Progress indicator

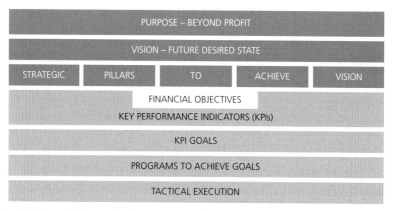

Fig. 3.4 The third step in the strategic planning process is complete – core focus areas in the form of strategic pillars have been developed and articulated in the form of a pillar mission statement. Collectively, these pillars also include all dimensions of Sustainability.

CHAPTER 4

Financial Objectives

Money was never a big motivator for me, except as a way to keep score. The real excitement is playing the game.

Donald Trump

For decades now, organizations have been obsessed with maximizing profit. Spurred by shareholders, investment firms, and bonus structures based on profit, executives are now more than ever focused on quarterly profits. This unhealthy obsession with short-term profit maximization has produced record profit margins but not without repercussions. Sure, organizations may enjoy some short-term profit increases but for the majority, these short-term gains come at the expense of longer term profit. Why? The answer is quite simple; a focus on profit takes away the focus from the real *profit drivers* such as value creation, employee engagement, research and development, innovation, community support, and so on.

At a high level, profit is really a factor of two variables: Revenue and Expenses. Improve one or both of these and improve profit, a fairly simple concept in theory. The rub is that in practice, it's much more complicated than it seems. It can be very difficult, at least in the short term, to predictably increase revenue. For this very reason, many managers and executives focus on reducing expenses. Whether

Purposely Profitable: Embedding Sustainability into the DNA of Food Processing and Other Businesses, First Edition. Brett Wills.

it is slashing employee wages, reducing headcounts, reducing capital or research and development (RnD) investment, these all result in immediate and predictable expense reductions and, therefore, profit increases – assuming Revenue stays relatively constant.

But for one reason or another, the majority of managers and organizations ignore the long-term repercussions of these short-term decisions. The repercussions are widespread and are not necessarily the focus of this book but, for example, reducing capital expenditures or investment in RnD means that future growth becomes much more difficult.

While finance is not the true focus of this book, the message here is that a focus on the *profit drivers* vs a short-term focus on profit itself, will in a majority of cases, produce more favorable results in the long term.

Let's not be naïve here, clearly profit is important and this should be a focus, but not the sole focus. As part of a robust strategy, organizations must pay attention to the financials and overall profit. The model presented throughout this book places profit as a key component to the overall strategy.

4.1 Understanding business objectives

The strategic planning model provided in this book requires that organizations have clearly defined financial or "business" objectives. These business objectives will be used to help guide the development of goals under each of the pillars developed in the previous chapter.

There are many approaches to setting financial or business objectives. Let's focus on a simple and straightforward approach for doing this. There are three high-level areas that smart organizations focus on to drive profit: Revenue, Gross Margin, and Overhead. Control these three things and control profit. Before getting into setting objectives for each of these, let's define each.

Most folks have a clear understanding of the first profit driver but for formality's sake, let's quickly review. *Revenue* is income that an organization receives from its normal business activities, usually from the sale of goods and/or services to customers. Some companies receive revenue from interest, royalties, or other fees. For non-profit organizations, annual revenue may be referred to as gross receipts. This revenue includes donations from individuals and corporations, support from government agencies, income from activities related to the organization's mission, and income from fundraising activities, membership dues, and financial investments.

Gross Margin represents the proportion of each dollar of revenue that the organization retains after paying the direct costs associated with producing the goods and/or services sold. Understanding Gross Margin requires an understanding of direct costs. Direct costs, also known as Cost of Goods Sold (COGS) is exactly as it sounds, the costs directly associated with making a product or providing a service. These costs typically include the cost of materials and the cost of direct labor. It does not include things such as sales or marketing costs, rent, insurance, administrative expenses, etc. Gross Margin is typically expressed as a percentage using the following formula:

Gross Margin(%) = (Revenue − Cost of Goods Sold)/Revenue

The level of Gross Margin will vary dramatically from industry to industry and even from company to company in the same industry. For example, food companies and airlines tend to have razor thin Gross Margins while software and consulting companies tend to have much fatter margins.

The amount of Gross Margin an organization has is then used to cover other indirect costs and expenses associated with running the organization. For example, if an organization sells a product for $10 and it costs $2.50 in materials and $2.50 in labor, the direct costs are $5. Using the formula above this organization would have a Gross Margin of 50% and therefore retains $5.00 from every product sold to contribute towards paying for things like rent, sales and marketing expenses, administrative support, and so on. These other expenses are known as Overhead, which we will look at next.

Overhead refers to all the ongoing, indirect expenses associated with running the business. They are expenditures which cannot be conveniently traced to or identified with any particular unit of product or service. In other words, they are necessary to the continued functioning of the business but the costs cannot be directly associated with the products or services being offered.

Overhead expenses can be fixed, meaning that they are the same from month to month such as rent, insurance, or management salaries. Or they can be variable, meaning that they increase or decrease depending on activity levels such as sales commissions. They can also be semi-variable, meaning that some portion of the expense will be incurred no matter what, and some portion depends on the level of business activity, such as energy or waste disposal costs.

Many organizations will express Overhead as a percentage of revenue. This is a good way to go about it as it allows for easier management as revenues fluctuate. Expressing Overhead as percentage of revenue is calculated using the following formula:

$$\text{Overhead \%} = \text{Overhead/Revenue}$$

So with the understanding of what is meant by Revenue, Gross Margin, and Overhead, it is easy to see how controlling these three things controls profit. Control the amount of income coming in and the direct costs associated with that income and then control all the other Overhead costs associated with supporting the sale of the products and/or services, and one controls the profit level.

Now, obviously there is much more to accounting and finance than these three high-level items. However, focusing on managing these three high-level items should be a priority as they are the major factors that produce profit. It is also important to note that this simplified approach works well for small to medium-sized enterprises (SMEs). Larger organizations, especially publicly traded ones, may choose to focus on different or additional financial measures. Regardless, organizations can think of these as their Key Financial Indicators (KFIs).

While there are many other financial metrics that can and should be measured, for most organizations, especially SMEs, these three indicators are the key ones to manage. To illustrate the process for setting business objectives, the three KFIs of Revenue, Gross Margin, and Overhead will be used. Organizations that choose to use different/additional KFIs will need to work through this process in the same way to set objectives for their own unique KFIs.

4.2 Setting business objectives

With the understanding that an organization must decide on their KFIs, it is time to set the financial objectives. This is an important, fundamental step in the overall strategic planning process as it will be used to align other goals later in the strategic planning process. Let's have a look at setting objectives for each of the three financial KPIs of Revenue, Gross Margin, and Overhead using the Smithville Foods example.

4.2.1 Setting a Revenue objective

In theory, setting a Revenue objective is rather simple. Pick a number and off to the races. Maybe it is a 10% increase year over year or maybe a fixed absolute amount such as $100,000. The challenge is picking the *right* number. Organizations must carefully consider where the increase in revenue is going to come from and if the increase is realistic. Typically, increases in revenue can come from two high-level areas: existing customers and new customers. Now, the driver for increases in these two areas could be from multiple sources.

Maybe the source is simply increased sales of existing products/ services to existing customers, or it may come from sales of existing products/services from new customers, or it may come from sales of new products to existing or new customers, or it may come from a price increase. Taking all these factors into consideration, the organization must determine a realistic objective for revenue that aligns with the end profit objective.

Using Smithville Foods as the example, let's see what this looks like:

Current revenue: $5,500,000.00
Existing customers: Increase by $400,000
New customers: $150,000
Revenue objective: Increase by 10% or $550,000 for a total of $6,050,000.

Based on current sales levels with existing customers, the sales team at Smithville determined that across all existing customers, a $400,000 increase is realistic if certain improvements are made and if scheduled new products are released. Furthermore, based on analysis of the market, the team feels that it is realistic to bring on two new customers at $75,000 each for a total of $150,000 if scheduled new products are released.

Organizations may have their own process for determining a realistic revenue objective, regardless of how the objective is arrived at; the point here is that a revenue objective must be set.

4.2.2 Setting a Gross Margin objective

Setting a Gross Margin objective can be a little more involved than setting a revenue objective because it is driven by multiple factors – all of the direct costs associated with producing the product or service. Setting the objective may be done in a few ways. The organization can identify opportunities and determine the cost savings, then use that to set the objective, or it may have the objective dictated because that is what is required to compete. Either way, the organization must review each of the high-level contributors to Gross Margin – Direct Labour and Materials – to identify the cost savings available.

Starting with Direct Labour, the organization must go through all the Direct Labour costs and determine if those costs can be improved. The goal here is not to lower people's wages or let people go but rather to determine if things can be done more effectively and efficiently. There are many ways to do this and a number of them are covered in more detail in the "Tools and resources" section near the end of this book. Some common examples include better training, increased engagement, and implementing continuous improvement programs such as Lean or Six Sigma, process innovation, and so on. The operations leader for the organization should have a good handle on the improvements available or at least a sense of how to identify them. Once the improvements have been monetized, the organization can use this to set a realistic but challenging objective for improving direct labour costs. For an objective dictated due to competition, going through this exercise will allow the organization to understand where they sit relative to the dictated objective and, if required, go back and look for further opportunities for savings.

A side note here – when volume is static, improving direct labour could end up creating excess labor for the organization. In this scenario, the first reaction is to let that extra labor go. Sometimes this is necessary; however, the smart organizations find ways to focus that extra labour on further improvements to create additional capacity and capability which supports and drives growth. In the situation where volume is expected to increase, the issue of excess labor no longer exists and the focus tends to be on holding direct labour constant with the higher volume.

Whatever the scenario, the organization must set the Direct Labor objective which could look something like the following example for Smithville Foods:

Current Direct Labour cost: $1,000,000
Direct Labour cost objective: Maintain @ $1,000,000 with 10% volume increase.
Cost avoidance: $100,000

Given that sales and therefore volume will increase by 10%, the operations leader feels that the current level of Direct Labour can be maintained through the improvements available. If the improvements were not made, the labour would rise directly with volume by 10%, so in other words, there is a $100,000 cost avoidance.

With the Direct Labour objective set, let's look at Materials. This is where it can get a bit tedious to determine a smart objective if one has not been dictated. The organization must go through each Material cost, line by line, and determine what improvements are available. This is a good time to pull in the operations leaders, procurement managers, and controllers. The improvements may come from improved buying practices, renegotiating contracts, redesigning products to use fewer materials, substituting materials or even small things such as lightweighting packaging, or implementing continuous improvement projects that reduce the amount of material that is being wasted. (See later "Tools and resources" section for support in identifying improvement opportunities.) Regardless of the solution, the organization will need to determine the improvements that are available for reducing the cost of Materials. Once this is determined, the improvements can be monetized to help set the material cost objective. Again, when an objective has been dictated, this exercise will let the organization know where it stands and, if required, it can go back and look for further savings. Continuing with the Smithville example, it could look something like this:

Current Material costs: $3,000,000
Forecasted Material cost: $3,300,000 (with 10% volume increase)
Material cost objective: Improve by 5% to $3,135,000

Due to the forecasted 10% increase in volume, Smithville first had to determine what the Material costs would be with the increased volume. Once this was determined, the operations team identified opportunities that totalled $165,000 or 5%. Some of the improvements identified were product redesign, improved buying practices, and specific lean initiatives as well as lightweighting some packaging.

With the objectives identified for the direct costs of labour and materials, the Gross Margin objective will need to be calculated. As seen earlier, this is done using the following formula:

$$\text{Gross Margin } (\%) = (\text{Revenue} - \text{Cost of Goods Sold})/\text{Revenue}$$

For Smithville Foods, it looks like the following:

Current Gross Margin: $\dfrac{\$5,500,000 - \$4,000,000(\$3M \text{ Materials} + \$1M \text{ Labour})}{\$5,500,000}$

$= 27.27\%$ or $\$1,500,000$

Margin objective: $\dfrac{\$6,050,000(10\% \text{ increase}) - \$4,135,000(\$3.135M + \$1M)}{\$6,050,000}$

$= 31.65\%$ or $\$1,915,000$

When Smithville calculated their Gross Margin objective, they used the numbers under a scenario of a 10% volume increase due to the 10% increase in sales. If there was no increase in sales, the revenue number would stay the same and the material and labor numbers would reflect the savings identified.

4.2.3 Setting an Overhead objective

Setting an objective for Overhead is fairly similar to setting an objective for Gross Margin. Overhead typically has more cost contributors than Gross Margin but the process for setting an objective is essentially the same. The organization must go through each cost contributor, line by line, and determine the improvements available. Since there are many fixed costs included in Overhead such as rent, insurance, and so on, some of the costs may prove difficult to improve over the short term. Other Overhead costs such as utilities, waste disposal, supplies, maintenance, etc., are easier to impact in the short term.

The improvements can come from energy conservation, water conservation, increasing waste diversion, improving maintenance practices, making better use of supplies, and the like. The opportunities will vary from organization to organization but the end goal here is to identify improvements in terms of cost savings available for each cost contributor under Overhead. With the savings identified, this can be used to help set an objective for Overhead. Like Gross Margin, Overhead is often expressed as a percentage of revenue using the following formula:

$$\text{Overhead \%} = \text{Overhead/Revenue}$$

For Smithville Foods, the Overhead objective looks like the following:

Current Overhead cost: $1,250,000

Current Overhead %: $\dfrac{\$1,250,000}{\$5,500,000} = 22.72\%$

Overhead cost objective: $1,227,350 ($22,650 savings)

Overhead % objective: $\dfrac{\$1,227,350}{\$6,050,000} = 20.29\%$

Smithville determined that Overhead would not be impacted by the 10% increase and also identified $22,650 worth of savings and used this to set the objective. If Overhead did increase with the increased volume, the net impact would have to be calculated and used to set the objective.

With the objectives set for Gross Margin and Overhead, let's take a look at the overall impact on profit:

	Current	New
Revenue	$5,500,000	$6,050,000
Direct Costs	$4,000,000	$4,135,000
Overhead	$1,250,000	$1,227,350
Profit	$250,000	$687,650

So there is a clear improvement in profit that comes from improving the three variables of revenue, Gross Margin, and Overhead; in the Smithville example it means an additional $437,650 in profit. Now remember, these are just examples and organizations may have their own way of setting financial objectives. Whether using this way or another, does not really matter, the point is this step in the strategic planning process requires that financial objectives be set. These objectives will guide setting goals under each of the Pillars.

At this stage in the game, organizations have a clear purpose for everything they do, have painted a clear picture of what the future will look like on the way to fulfilling the purpose, have identified the focus areas for moving towards the desired future state, and set the financial objectives that will be realized in the short term. Figuring out a path for doing this is the next phase in the strategic planning process, starting with developing a clear set of KPIs that will gauge performance under each pillar. Then setting goals for each of those KPIs that, when achieved, will align back to achieving the financial goals. Following this, the organization must then figure out what programs and initiatives will be worked on to achieve the goals and then how to tactically execute the overall strategy.

Progress indicator

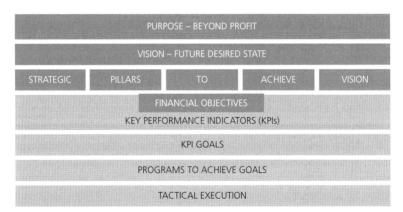

Fig. 4.1 The fourth step in the strategic planning process is complete. Financial objectives have been set.

CHAPTER 5

Measuring What Matters – KPI Development

Not everything that can be counted counts, and not everything that counts can be counted.

Albert Einstein

Anybody who has been in the corporate world for more than 5 minutes has heard the old management saying by Peter Drucker: "What gets measured, gets managed." This is true, so be careful what gets measured. For whatever reason, people love to measure things, especially in the corporate world. If a measure can be slapped on it, then it probably will. Maybe it is because measurement provides a sense of accomplishment or purpose but measurement is powerful and when pointed in the wrong direction, it can be powerfully disrupting.

If the wrong stuff gets measured, then the wrong stuff gets managed. Organizations do this constantly and then busy themselves managing all kinds of stuff that doesn't really matter. The result is that people lose focus on the things that drive value and end up focusing on other things that do not necessarily bring real value to the organization. Additionally, measuring the wrong stuff sucks up resources, time, capital, etc., because the act of measurement alone can be a tedious and complex endeavor that requires people's time, maybe

Purposely Profitable: Embedding Sustainability into the DNA of Food Processing and Other Businesses, First Edition. Brett Wills.
© 2016 John Wiley & Sons, Ltd. Published 2016 by John Wiley & Sons, Ltd.

investment in software systems, and development of policies, procedures, and so on.

There are some classical examples of mismeasurement. For example, many companies like to measure the training provided to employees. Great intention, but measuring the number of training hours drives organizations to simply focus on providing more training regardless of the value it provides. This causes organizations to devote resources to developing training programs, delivering the training, counting the training hours for each employee, and so on. Instead, organizations could measure the impact training is designed to have, such as reduction in defects or reduction in injuries, increased productivity, and so on.

With an understanding that measuring the wrong stuff can be detrimental to organizations, it is also important to understand that measuring the right stuff can be of great value.

To start, measuring the right stuff makes it easier to set clear and meaningful goals, it provides a mechanism to build accountability, and like Peter Drucker says, it drives action. There are many other benefits associated with measuring the right stuff, such as providing signals to tell the organization when it is doing the right things or adversely, doing the wrong things. In short, measurement is an important and valuable activity for organizations. However, one must always remember to measure what matters and to recognize that sometimes things can't be measured, so don't try to put a yard stick to something just for the sake of it.

5.1 Understanding KPIs and metrics

There is much confusion around measurement. Especially when it comes to measuring organizational performance at a high level, these types of measurements are often referred to as key performance indicators or "KPIs". One major misnomer is that a KPI is an action to produce a result vs measurement of what results from an action. It is an indication of performance not the action that will produce the performance. Another point of confusion around KPIs is the difference between KPIs and metrics. Many folks think they are one and the same, others recognize there is a difference but do not really understand what the difference is. Key Performance Indicators are exactly that, *key* performance indicators. They measure factors that are critical to the success of the organization at a high level. Organizational performance, and thus profit, is directly related to the direction of these indicators; if they improve, organizational performance improves and vice versa.

Metrics on the other hand, are performance indicators that feed into or impact the KPIs. While a change in the direction of a metric may have some impact on the organization and may be important, they are not necessarily critical to organizational success. For example, take something like customer loyalty. A KPI should measure how loyal customers are so a KPI in this area may be something like "repeat orders" or "actual vs projected sales." A metric or performance indicator would then be things that influence loyalty such as the results of a customer loyalty survey. The survey score would identify the areas that are driving or influencing customer loyalty and enable actions to be taken in order to improve loyalty and thus improve the customer loyalty KPI. So in other words, KPIs are higher level performance indicators that are critical to success whereas metrics are lower level measures that impact higher level KPIs. One way to better understand this is to understand the difference between leading and lagging indicators.

5.1.1 Leading vs lagging

There is much talk in the management world about leading and lagging indicators but exactly what does this mean? Lagging indicators tend to be more *output* oriented where leading indicators tend to be more *input* oriented. Both are valuable but serve different roles. Lagging indicators provide value in that they measure results; however, they do not necessarily drive action to produce the results. Leading indicators on the other hand do not necessarily measure results per se but rather measure the effectiveness of activities that will drive action towards the results.

A common example used to explain leading vs lagging indicators is weight loss. The lagging indicator is the amount of weight that has been lost (or gained) and the leading indicators are caloric intake and calories burned. In this example, weight loss would be the KPI, where caloric intake and calories burned would be metrics.

Parlaying this to the business world, many companies like to measure "On-Time Delivery". This can be viewed as a lagging indicator as it measures the outcome of whether orders were delivered on time or not. It is valuable information to know on-time delivery rates but the horse is already out of the gate and therefore a lagging indicator. Leading indicators in this area could be something like the number of material or production delays. They are leading because they influence delivery and provide an indication of what the on-time delivery performance may look like. They also allow something to be done today to influence the delivery. For example, if there are some considerable material delays, chances are the order may be late, but if identified ahead of time something can be done to remedy the problem such as expediting material delivery or scheduling production such that extra resources can be assigned to the order once the materials arrive to get it out in time.

In summary, KPIs are more of lagging indicators as they measure output and results, where "metrics" should be more leading indicators as they measure input and actions required to produce the desired output. But remember that KPIs are indicators that are critical to success so just because it is a leading indicator does not mean that it cannot be a KPI.

Beyond leading and lagging, both KPIs and metrics can be expressed in different ways; let's have a look at the 2 major ways.

5.1.2 Absolute vs normalized measures

When choosing a measure it is important to understand the difference between Absolute and Normalized or "Intensity" based indicators. Absolute indicators measure the net impact and are usually expressed as a whole number. Normalized or intensity based indicators measure relative impact and are usually expressed as a ratio by dividing the absolute indicator by some denominator such as units of output or number of employees. For example, many companies these days are measuring their carbon footprint and setting carbon reduction goals. Using the absolute indicator of tonnes of carbon dioxide equivalent (CO_2e) measures the net increase or decrease in carbon emissions. Using a normalized indicator such tonnes of CO_2e per unit of output, measures the relative increase or decrease based on units of output produced. An organization can improve (or decrease) their normalized footprint while seeing an increase in absolute emissions – total emissions went up but when spread out over the increase in units produced, the emissions per unit actually went down.

There are pros and cons for both Absolute and Normalized Indicators. Absolute indicators measure net impact and therefore when applying absolute indicators to activities that vary based on volumes, the organization is penalized for business growth or rewarded for business decline – sometimes this is a good thing, other times it may not be. Normalized indicators measure impact relative to volume or activity and therefore when applying normalized indicators to activities based on volumes, the organization receives a truer picture of performance relative to volumes – again, sometimes this is a good thing and sometimes not.

Take the above example of carbon footprinting. Because carbon emissions can be heavily influenced by volumes, using an absolute indicator will reward the organization for business decline and penalize them for business growth. A normalized indicator will provide a better indicator of true performance because emissions are spread out over units of output and while overall emissions will go up, the amount of emissions generated per unit of output typically goes down. For this reason, many organizations use normalized indicators for activities such as carbon emissions, water consumption, and so on.

Take the earlier example of on-time deliveries as another scenario. In this case, business volumes do not really matter as the organization is trying to determine, regardless of business volumes, whether orders are delivered on time or not.

At the end of the day, it is important to understand the difference between normalized and absolute indicators and choose the best type for the particular situation. In general, absolute indicators are used when organizations are trying to measure net impact regardless of business volumes or want the measure to include fluctuations in volumes such as customer loyalty. Normalized indicators are typically used when organizations want to measure outcomes relative to business volumes, which many organizations do with carbon.

5.2 Pillar KPI development

When developing KPIs it is clearly important to measure what matters and a good rule of thumb when doing this is to not pick from a laundry list of KPIs. It is important to understand and be aware of the different KPIs available but just because other organizations measure certain things does not necessarily mean that those are the right things for every organization – or even similar organizations – to measure. Granted, there are certain metrics that tend to be standard across industries and business in general, such as safety KPIs. Additionally, some organizations may decide to measure some things that others measure, but they should not just jump to copying what others are doing. Organizations must develop their own unique set of KPIs if they are going to be successful in measuring the right stuff.

The good news is that by following this approach, organizations have already identified what is important to them through the pillars and corresponding pillar statements. Let's work through the process for developing a meaningful set of KPIs using Smithville Foods' pillars and pillar mission statements as an example.

Step 1: Identify what needs to be measured

The first step in developing meaningful KPIs is to identify what actually needs to be measured. For many organizations, this can be a difficult process simply because they have not identified what is important to them (the factors that are critical to their success aka Strategic Pillars). By following the strategic planning system laid out in this book and identifying pillars of focus with a corresponding pillar mission statement, organizations have already decided what is important to them and, aspirationally, what they are trying to achieve under each pillar (in other words, the factors critical to their success).

Therefore, because KPIs are defined as measures which are critical to success, and the organizations pillar statements identify those factors critical to success, an organizations KPIs then become the measures that gauge performance under each pillar.

To begin developing KPIs, organizations must go through the pillars one at a time and determine what needs to be measured based on the pillar mission statement. This is done by picking out the key actions or results listed in the pillar mission statement. At this point, do not worry about metrics vs KPIs, leading vs lagging, or absolute vs normalized, or how many things are being measured; this will all be dealt with in subsequent steps.

Pillar: Customers
Pillar mission statement: *Build loyal partnerships with our customers through a focus on radical transparency and nimble customer service.*
Measurable(s): Loyal customers, Transparency, Nimble customer service

In this situation it seems fairly apparent that "loyal customers" is the main measurable since the action is to build loyal partnerships with customers. Transparency and nimble customer service seem more like the way to achieve loyalty and thus may be more supporting metrics than a key measurable outcome. Nonetheless, at this stage, the point is to identify all possible measurables, so all should be noted. This type of thinking should be used when working through the rest of this step for the remaining pillars – capture all possible measures that could gauge progress under the pillar. For some pillars, there could be multiple possible measures and others there may only be one or two.

Pillar: Products
Pillar mission statement: *Offer quality, innovative products that are soft on the environment and provide social benefit all at a fair price.*
Key Measurable(s): Quality products, Innovative products, Environmental impact of products, Social impact of products, Price of products

Pillar: Operational excellence
Pillar mission statement: *Efficiently provide a stable supply of quality food products through a focus on innovative processes, continuous improvement and environmental sustainability.*
Measurable(s): Stable supply, Quality, Process innovation, Continuous improvement, Environmental impact

Pillar: People
Pillar mission statement: *Foster a safe workplace full of engaged and agile employees who are motivated to achieve our vision.*
Measurable(s): Employee safety, Employee engagement, Employee agility, Employee motivation

---

Pillar: Resilience
Pillar mission statement: *Develop decentralized systems that mitigate risk by quickly responding to business interruptions.*
Measurable(s): Decentralized systems, Risk mitigation, Response to business interruptions

---

At this point, organizations will have a good list of the things that could be measured in order to gauge performance under each pillar. However, at this point, there may be quite a large number of measurables identified. Do not worry about that at this point; distilling down to a few key measurables will be dealt with in the next step.

Step 2: Identifying KPIs vs metrics

This step is where the measures can be further distilled down to identify the key measurables which will be the basis for developing the KPI (it is not technically a KPI until it is expressed numerically). In order to identify the key measurables aka KPIs, the supporting measurables (aka metrics) need to be sifted out. This is more art than it is science but remember that KPIs are critical to performance and tend to be lagging indicators as they are trying to measure an outcome vs inputs that will drive or contribute to an outcome. Here is some guidance to help sift out the supporting metrics to determine the KPIs:

- Does the pillar statement provide any hints? For example, the statement may say to "achieve x" by or through "y" and/or "z" such as the customer pillar statement Smithville – "build loyal partnerships … through a focus on radical transparency and nimble customer service." If so, the measurables that measure y and/or z are probably the supporting metrics with x being the KPI. In this case, Loyal Partnerships is the KPI (x) with transparency (y) and nimble customer service (z) being the supporting metrics.
- Do any of the measures influence the outcome of another measure? If so, then the indicator that influences is leading and probably a metric.
- Are any of the measures measuring the same thing? If so, pick the best one.

- Can it be accurately measured? If not, strongly consider if it is possible to develop a system to measure it and if not, maybe it can't be a measure.

Let's continue with the Smithville Foods example to work through this step in the process:

Pillar: Customers
Pillar mission statement: *Build loyal partnerships with our customers through a focus on radical transparency and nimble customer service.*
Measurable(s): Loyal customers, Transparency, Nimble customer service
KPI: Customer loyalty
Supporting metrics: Transparency, Nimble customer service.

Loyal customers is a potentially good KPI because it is more of a lagging indicator in that it gauges an organization's performance when it comes to customer loyalty. However, "transparency" and "nimble customer performance" are more leading indicators to loyalty as they contribute and influence how loyal customers are, therefore they make potentially good supporting metrics.

Pillar: Products
Pillar Mission Statement: *Offer quality, innovative products that are soft on the environment and provide social benefit all at a fair price.*
Measurable(s): Quality products, Innovative products, Environmental impact of products, Social impact of products, Price of products
KPI: Social impact, Environmental impact, Innovative products, Quality products
Supporting metrics: Price of products

This is a tough one. At first glance they all seem to be measuring an outcome, however, this is not the case. Take the price of products for example, this is actually trying to measure if the price is fair or not – this can be tough. Who determines what fair is? For this reason, while it is important to offer products at a fair price, it will be hard to measure and it is rather subjective so this may not be a good KPI; however it may be considered as a supporting metric if a means of measurement can be determined. Social and environmental impacts can be measured rather effectively today and they are also outcomes of the product design, delivery, use, etc., so it makes them a potentially good

KPI. An innovative product can be hard to measure but innovation in itself can be measured in certain ways such as – percentage of revenue from products less than 2 years old. Since this is the measure of an outcome of innovation, it may also be a good KPI. Quality can certainly be measured and is definitely an outcome of performance, so it looks like this pillar has a high number of KPIs.

Pillar: Operational excellence
Pillar mission statement: *Efficiently provide a stable supply of quality food products through a focus on innovative processes, continuous improvement and environmental sustainability.*
Measurable(s): Efficiency, Stable supply, Quality, Process innovation, Continuous improvement, Environmental impact
KPI: Efficiency, Quality
Supporting metrics: Stable supply, Process innovation, Continuous improvement, Environmental impact

---

Looking at the measurables and the statement for this pillar, the outcome to be measured here is "efficiency," "stable supply," and "quality food products." However, it can be argued that stable supply is more of an influence on efficiency since it is hard to be efficient without a stable supply and therefore may make a better supporting metric to still help drive a focus on responsible sourcing – a key ingredient for a stable supply chain. Innovative processes, continuous improvement, and environmental sustainability are all contributing factors to the outcomes of these three things and therefore make good supporting metrics.

Pillar: People
Pillar mission statement: *Foster a safe workplace full of engaged and agile employees who are motivated to achieve our vision.*
Measurable(s): Employee safety, Employee engagement, Employee agility, Employee motivation
KPI: Employee safety, Employee engagement
Supporting metrics: Employee motivation

Safety is definitely an outcome and would make a good KPI. Typically a motivated employee is an engaged one and therefore, employee engagement would be an outcome of motivation (and other things) and thus engagement would make a good KPI whereas motivation would be more of a supporting metric. However, employees may be engaged but not agile, but agility is a tough one to measure and so while in theory it make may make a good KPI, it may not be practical – so something to keep in mind but not a KPI for now.

Pillar: Resilience

Pillar mission statement: *Develop decentralized systems that mitigate risk by quickly responding to business interruptions.*

Measurable(s): Decentralized systems, Risk mitigation, Response to interruptions

KPI: Response to business disruptions

Supporting metrics: Decentralized systems

At first glance it may seem like the main outcome being aspired to in this pillar is "mitigating risk." However, measuring risk is very difficult – measuring financial risk can be done but overall business risk can be challenging to measure since there are so many sources of risk. For this reason it may not make sense to have an overall KPI for risk since one number wouldn't be very indicative of performance. However, measuring if there are decentralized systems in place and more importantly measuring response to business interruptions as a result of effective systems being in place does make sense. Since the systems contribute to the response time it would then suggest that response time is the KPI and decentralized systems be a supporting metric.

As stated earlier, this step is not science; the purpose of this exercise is to get a feel for how to differentiate between KPIs and metrics and the process for doing so. The best way to do this is to have a solid, fundamental understanding of the difference between KPIs and metrics.

Also, remember that at this point in the strategic planning process, the focus is on identifying the measures that are critical to success which are the KPIs, not the metrics. The supporting metrics will come into play later in the strategic planning process when it comes to developing the programs and initiatives to improve the KPIs – so put the supporting metrics aside for now and focus on the KPIs.

Step 3: Defining the KPI number

With the KPIs determined, the actual number used to measure the KPI needs to be defined. The rub here is that many times there are multiple ways to measure the same thing. Take for example, customer loyalty. This can be measured in a number of different ways. It could be through a customer loyalty survey, repeat purchases, sales vs previous period, and so on. The organization must determine the best approach for their own unique situation. Additionally, this is where the organization must determine if it is going to use an absolute or a normalized indicator. Continuing with the Smithville Foods example, let's work through this process.

Pillar: Customers
KPI: Customer loyalty
KPI number: Sales vs previous period (expressed as %)

As discussed, customer loyalty can measured in a number of ways. However, the real reason for focusing on customer loyalty is to drive sales, so it only makes sense to use that measure for the actual KPI number.

Pillar: Products
KPIs: Social impact, Environmental impact, Innovative products, Quality products
KPI number (Social impact): No. of meals provided from money donated.
KPI number (Environmental impact): Life cycle assessment (LCA) score
KPI number (Innovative products): % of sales from products 2 years old or less
KPI number (Quality products): No. of defective products to market/no. of products shipped

---

There are multiple KPIs under this pillar so let's look at each in turn. For social impact, it is really trying to measure how the social efforts impact the health of society; however, this can be extremely difficult and tedious to measure. Therefore, with an understanding that the more meals provided, the better off society will be, the number of meals provided would make a good KPI number here. When it comes to environmental impact, there are many things such as carbon impacts, water impacts, biodiversity impacts, etc., that can be measured. Since an LCA measures a large variety of impacts, it makes a good KPI. Other organizations may choose to focus on a few specific impact areas and measure those. For innovative products, the main focus here is really using innovation as a means to drive further sales, therefore it makes perfect sense to measure the sales that result from innovation. Quality is a rather simple one but it can be measured in many ways depending on what the focus for quality is. In this case quality is focused on preventing defective products from hitting the marketplace, so the above measure being used by Smithville of defective products to market would work well. Another measure of quality that is more internally focused will be explored in the next pillar.

Pillar: Operational excellence
KPIs: Efficiency, Quality
KPI number (Efficiency): Units produced/maximum output
KPI number (Quality): No. of defects/units produced

For the operational excellence pillar, the two key measurables here are efficiency and quality. These are rather standard measures but again can be measured in different ways. For efficiency, the actual number used to measure will depend on how the organization defines efficiency. Usually it is some sort of measure of current output vs ideal or maximum output as shown by the Smithville example. Quality in this instance is really focused more on internal quality to avoid rework and therefore the number here of defective units relative to total units produced reflects that view.

Pillar: People
KPIs: Employee Safety, Employee Engagement
KPI number (Employee safety): Total recordable injury rate
KPI number (Employee engagement): Gallup Q12 survey score

Employee safety and employee engagement are both common measures for all types of organizations. Furthermore, there are rather standardized ways to measure these areas such as the ways noted in the Smithville Foods example above. However, organizations may have their own unique way of measuring these areas, which is also fine as long as the number focuses on performance results or output vs input. For example, measuring the amount of safety training hours or safety meetings contributes to better safety performance but does not measure the actual performance output or results and therefore would not really be a good KPI.

PIllar: Resilience
KPI: Response to business disruptions
KPI number: Dollars lost due to business disruptions

Under this pillar, the focus is really on minimizing the financial impact to the business from disruptions, therefore, the measure of lost dollars aligns with the intent of this pillar.

Ok, so at this point the organization should not only understand what needs to be measured but also have a clear idea of the actual number that will be used to measure it. The work on KPIs is not finished yet though; there are still a few more steps to be completed before closing off this section of the strategic planning process.

Step 4: Building the baselines

Now it is time to build on the foundational work of identifying the KPIs to build the baselines for each KPI that will be the trigger for beginning to drive action. This is a rather straightforward step for smaller organizations but like many things, can get more complicated the larger the organization. The first thing the organization will need to do in building the baseline is determine the base year for KPIs. The base year serves as the starting point for which to judge performance over time. Since the KPIs are designed to measure performance under the new strategic plan, many organizations may choose to simply set their base year to match the year they launch the new strategy. This approach works but organizations may want to consider going back at least 1 or 2 years in order to get a "before & after" perspective.

Once the base year has been determined, the organization can then begin building the actual baselines. Depending on the size and complexity of the organization, this may be a rather simple step of designing a process to collect, enter and calculate the data or it could be more complicated if it requires the use of software or involves collecting data from multiple divisions, facilities etc. Either way, the organization will need to complete the following actions to build the KPI baselines:

1 Determine if the data is currently available. If yes, move on to action number 2. If data is not available, the organization will need to set up a system to begin collecting the data.
2 Determine the frequency at which the data will be collected – i.e weekly/monthly/quarterly/annually, etc. Note here that the frequency does not have to be consistent across each KPI. Some KPIs may be updated more frequently than others.
3 Set up the processes and systems required to collect, manage, and report the data.
4 Assign responsibility for collecting, managing, and reporting the data. Responsibility may vary for the different KPIs.
5 Pull the data for each KPI starting with the base year. Complete the calculations required for each KPI.
6 With calculations completed, create a chart/graph for each KPI that will be used to analyze performance over time.

Continuing with the ongoing example of Smithville Foods, the KPIs for Smithville will now look something like those shown in Table 5.1:

Table 5.1 KPI description, number and current status for each pillar.

Pillar	KPI description	KPI number	Current KPI status
Customers	Customer loyalty	Sales vs previous period (%)	Flat @ 0%
Products	Social impact	No. of meals provided as a result of money donated from sale of products	1,297
	Environmental impact	LCA score	17
	Innovative products	% of sales from products 2 years or less	6.5%
	Quality products	No. of defective products to market/ no. of products shipped	1.2%
Operational Excellence	Efficiency	Units produced/maximum output	78%
	Quality	No. of defects/units produced	2.1%
People	Employee safety	Total recordable injury rate	18.3%
	Employee engagement	Gallup Q12 survey score	3.3
Resilience	Response to business disruptions	Dollars lost due to business disruptions	$100,000

At this point, the development of KPIs to measure performance under each pillar is complete. In the above example, for brevity sake, the KPIs are only showing the current state and only presented in a table format. As discussed earlier, it is worth showing historical performance of each KPI in some graphical format.

It is important to reiterate that the example used here is simply to illustrate the process for developing a unique set of KPIs and are not suggesting that organizations use the metrics showcased but rather work through the process to develop their own customized set of KPIs that will measure performance of each pillar aka – the factors critical to their success.

5.3 Building a dashboard

With the KPIs developed, organizations may choose to build a dashboard that can be used to report out the KPIs. A dashboard is simply a one page summary of the current status for each KPI. A dashboard can be a simple table or it can get fancier through the use of graphics or visuals. Using the earlier example of Smithville Foods, a simple dashboard would look something like the following:

KPI dashboard					
Pillar	KPI Description	KPI number	Current	Goal	Variance
Customers	Customer loyalty	Sales vs previous period (%)	0%		
Products	Social impact	Meals provided	1297		
	Environmental impact	Lca score	17		
	Innovative products	% of sales < 2 years	6.50%		
	Quality products	Defects / products shipped	1.20%		
Operational excellence	Efficiency	Units produced / max output	78%		
	Quality	No. of defects / units produced	2.10%		
People	Employee safety	TRIR rate	18.30%		
	Employee engagement	Q 12 score	3.30%		
Resilience	Response to business disruptions	Dollars lost to business disruptions	$100,000		

There are two empty columns in this dashboard – one for the goal and one for the variance or "gap" between the current status and the goal. These will be addressed in the next steps of the strategic planning process and can be filled in at that time – leave them blank for now.

A special note here. This tends to be the spot where organizations begin to ask about cascading the strategic plan. Depending on the size of the organization, cascading of the KPIs to different facilities, divisions, etc., may be required. However, the entire strategic plan must be cascaded in order to facilitate proper alignment across the organization. Cascading the strategic plan is typically required in larger, more complex organizations with more than one facility, division, etc. For smaller organizations, cascading may still be required. However, each piece of the strategic plan is cascaded differently based on the complexity of the organization. Even when cascading is required, there must be a core strategic plan in place to cascade down, so cascading will be covered in Chapter 8, which focuses on tactical execution.

The KPI development step of the strategic planning process is now complete and the organization will now have developed a meaningful set of KPIs that will measure the elements critical to organizational success as defined by the pillars and pillar statements. As one can see, the process is a little more involved than just simply choosing some standard KPIs but there is tremendous value in following this approach. The organization will now be focused on managing the areas that will add value and drive organizational performance – in other words they will be focused on the "right stuff." To boot, the organization also has a means to measure sustainability performance since the collective KPIs measure social, environmental, and economic performance. The supporting metrics identified will also contribute to measuring sustainability performance and together with the KPIs, the organization will be able to transparently report sustainability performance to stakeholders. The next piece of the strategic planning process is building on the work completed so far to develop clear goals for each of the KPIs.

Progress indicator

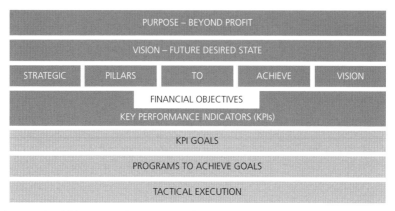

Fig. 5.1 The fifth step in the strategic planning process is complete – KPIs for measuring performance under each pillar have been identified and the baselines for each KPI have been established.

CHAPTER 6

Setting Expectations – KPI Goal Development

Setting goals is the first step in turning the invisible into the visible.

Tony Robbins

Goals are important. Give someone a bow and arrow, tell them to shoot and the response will likely be "At what?" When there is nothing to shoot at, there is no reason for taking a shot and if a shot *is* taken, the arrow ends up where the arrow ends up. Provide something to shoot at and everything changes. All of a sudden there is a reason for shooting, an end goal in mind, a challenge to meet, something to measure progress and a direction to point efforts.

There are countless surveys and studies demonstrating the power of goals. One of the most well-known studies comes from Harvard (McCormack, 1986). In 1979 interviewers asked new graduates from the Harvard MBA program a simple question: "Have you set clear, written goals for your future and made plans to accomplish them?" The interviewers found that:

- 84% of graduates had no specific goals
- 13% had goals but they were not committed to paper
- 3% had clear, written goals and plans to accomplish them.

Purposely Profitable: Embedding Sustainability into the DNA of Food Processing and Other Businesses, First Edition. Brett Wills.
© 2016 John Wiley & Sons, Ltd. Published 2016 by John Wiley & Sons, Ltd.

Ten years later in 1989, the graduates were interviewed again. The results were clear when the interviewers found that:

- The 13% who had goals were earning, on average, twice the amount of the 84% who had no specific goals.
- Even more interesting, the 3% who had clear, written goals, and plans were earning, on average, ten times more than the other 97% combined.

This is just one of many studies that reveals the power of setting clear goals to drive performance. The kicker here is that the power of goals applies to both personal performance and organizational performance. Given that organizations are really a team of "people," this makes sense. Give the organization something to shoot at and it changes everything. Each of the people who comprise the organization now have something to shoot at and a common goal to work towards. Successful organizations have long understood this power and there are many studies illustrating the dramatic performance increases that are realized when organizations set clear goals.

Why is it that organizations who set clear goals, outperform those who do not? There are many reasons. From providing clear direction and harnessing employee energy and creativity to clarifying expectations and providing a mechanism for accountability – goals are a critical ingredient to driving organizational performance. The trick is setting the right goals that are properly aligned.

Some folks say the right goal or a good goal should be SMART – Specific, Measurable, Attainable, Realistic, and Timely (Doran, 1981). This is great, goals *should be* smart and a good goal should have all these characteristics. However, there is one major piece missing from this equation: *alignment.* A goal can have all the characteristics mentioned but if the goal itself is not aligned then it is not a very "smart" goal. There are two points of alignment for goals: alignment to elements that are critical to success and alignment to financial objectives. First, the goal must be aligned to drive the elements that are critical to success. For example, if an organization sets a "SMART" goal for improving on-time delivery and on-time delivery is not a critical component to success, than it doesn't matter how specific, measurable, attainable, realistic, or timely it is. It is not aligned and, therefore, not very smart as it drives efforts towards areas that are not critical to success and does not bring the organization closer to achieving the longer term vision.

In order to align the goal with driving the areas that are critical to success, organizations must first clearly define the areas that are critical to success, have a set of KPIs that measure this success then set goals to improve those KPIs. Now, defining the areas that are critical to success requires having a vision of the future. Having a vision of the future requires a longer term purpose for achieving that vision. Fortunately, by following the system laid out in this book, all these things have been completed in the right order and therefore the first point of alignment has already been connected.

The second point of alignment for goals requires that they realize the financial business objectives that were set in Chapter 4. This is a critical piece of creating alignment and if not completed, the goals – and therefore the rest of the strategic plan – will not be aligned. Many organizations make this mistake and set goals for KPIs that do not align with their financial goals. For example, let's say an organization has a financial objective to reduce overhead by $50,000. If the cumulative savings realized from achieving the goals for each of the KPIs that contribute to reducing overhead is only $30,000 then there is a problem. The cumulative savings realized from achieving the goals does not total the $50,000 goal. The same thing goes for gross margin and revenue objectives. In order to facilitate this, organizations must understand how much money will be saved or generated through achieving the goals that are set under each pillar.

As one can see, setting a goal for the KPIs that connects with the financial objectives creates an aligned goal but, furthermore, by following this process and setting a goal for the KPIs, it also makes the goal specific and measurable (The "S" and "M" in SMART) since the KPIs are specific and measurable. By completing this step in the process to set goals for these KPIs that are attainable, realistic, and timely, it will produce a truly "smart" goal. These goals will serve as the catalyst for driving action – action in the form of programs that will guide day-to-day activities of the organization – more on this in the next chapter.

A side note to consider here. It is important to pay attention to the language used when talking about goals. There are many words used to describe goals such as objectives, targets, aspirations, and so forth. The language used to communicate "goals" can have a strong impact on how they are perceived, the mindset they are approached with, and ultimately influence whether they are achieved or not. For example, referring to a goal as an objective makes it sound soft and optional. What others may hear is "let's try and do our best to achieve this." Likewise, referring to a goal as a target suggests that initially it will be missed.

Referring to a goal as an aspiration insinuates that it won't be achieved today but maybe over time it will be realized. A goal should be strong and certain, failure is not an option, it must be realized. Even using the word "goal" could be taken by some as the ideal end result. For these reasons, organizations may consider changing the language used when talking about goals and refer to them as "commitments." A commitment communicates that there is no option for failure, there is a clear duty to achieve the end result.

To aid in developing a smart goal or "commitment" it helps to have an understanding of the different types of goals. Let's look at the major different "types" of goals.

Short term vs long term: Goals can be set for the short term or long term. There are many views on what is considered short term vs long term but generally speaking, short-term goals are less than 2 years and long-term goals reach out beyond 2 years. Organizations can also set what are known as "stretch" goals. These goals reach out a number of years, say 4, 5, or more years into the future. They are referred to as stretch goals because they "stretch" out what can be realistically achieved within the next few years. A combination of both long-term and short-term goals can also be set but should line up with each other. For example, there could be a long-term goal of improving revenues by 25% over 5 years with a 5% improvement year over year.

Absolute vs Normalized: Like with KPIs, goals can be absolute or normalized. Typically, determining whether a goal is absolute or normalized is dictated by the KPI the goal is being set for. If the indicator is a normalized indicator, then the goal will be normalized and vice versa.

Hard vs Soft: In addition to long term and short term, absolute and normalized, goals can also be either hard or soft. Hard goals are sharply defined, set out clear improvements, and most notably prescribe a specific number improvement such as a 5% reduction or a 10% increase. Soft goals are also sharply defined and set out clear improvements but do not prescribe a specific number improvement but rather prescribe a specific outcome such as "complete this" or "develop that." In most cases, soft goals should be short term or temporary and ultimately lead to setting hard goals. Furthermore, certain types of soft goals should be avoided. For example, setting a soft goal to "improve customer service" leaves a cloudy picture of what is to be achieved and it makes it hard to hold people accountable since most people will have a different idea of what "improve" means. This type of soft goal should be avoided. On the other hand, setting a soft goal to establish a hard goal or setting a soft goal to develop a baseline so that a hard goal can be set, makes sense but should only be a means for landing on a hard goal.

While this is not an exhaustive list of all the types of goals per se, they fundamentally cover the different types of goals aka "commitments" that can be set by an organization. Also, there is not necessarily a right or wrong "type" of goal to use; it all depends on the unique situation of the organization. For example, if an organization has more absolute than normalized KPIs, they will set more absolute than normalized commitments. Typically, an organization will have a combination of different types of goals that best suit their needs. At the end of the day, the important thing to remember here is that, sure a goal needs to be the right "type" of goal but, more importantly, the goal must be a smart goal as discussed earlier.

Before setting the actual commitments for the KPIs that have been put in place, there is one last area to look at. A common challenge when going through this step in the process is determining what level the goal should actually be set at. Should it be 10%, 5%, 20%? A common answer to this is that it should be "SMART" or it should be realistic but challenging. Great, but that doesn't answer the specific question of what level it should be set at. What is realistic? What is challenging? Well, there are a few ways to figure out the level at which a goal should be set at.

Goals are too important to take an ad hoc approach for establishing the level at which it will be set. In order to ensure a goal is truly a "smart" one, a systematic, almost scientific approach must be taken for establishing the level at which the commitment will be set to ensure it is not only realistic and attainable but also challenging. At the same time, excessive analysis can paralyze the process so the approach must be straightforward. Let's look at three straightforward ways for determining the proper level to set a goal at.

External benchmarking

A popular way to determine the level at which a commitment should be set is through external benchmarking. This method analyzes what other organizations are doing, the level others have set goals at, and the performance achieved to provide a point of reference for comparative analysis. Organizations may look at similar or like organizations within their industry or go outside their industry and look at overall best in class benchmarks for a particular area. While in theory external benchmarking is a fairly simple concept of researching what others are doing, in practice it can be much more difficult. Even though more and more organizations are becoming increasingly transparent, it can still be difficult to find the specific data and information required. Beyond this, even if the data and/or information is available, many times the context in which it was defined is not available and therefore, may be misleading. Having said that, external benchmarking provides a great jumping off point for discussing what level a commitment should be set at and – when the correct information is available – provides a great reference for determining the level at which to set a commitment.

Internal benchmarking

Another way to set a level of commitment is through internal benchmarking. Some managers refer to this as "racking and stacking" where an organization's different departments or facilities are ranked relative to each other. The top performing departments or facilities demonstrate what is possible and this performance becomes the level at which goals are set for the other departments or facilities. This works well in situations where the comparison is between similar areas, departments, or facilities. It becomes a little more challenging when the departments or facilities being compared are dissimilar. A pitfall of this approach is limiting performance to what has been achieved internally vs the potentially greater results being realized outside the organization. This approach is a good place to start for organizations who are beginning the journey or refocusing their efforts.

Once all departments, facilities, etc., are firing at equal levels, other approaches can be used to elevate the levels at which commitments are set. Even for small organizations with only one facility, this approach can be taken by looking at historically high results.

Opportunity based benchmarking

While this approach is also focused internally, opportunity based benchmarking is different than internal benchmarking. Opportunity based benchmarking looks at internal improvement opportunities.

The goal or commitment is then set at the level where performance would be if those opportunities or improvement were actually realized. For example, if an audit or analysis identified that a 10% improvement was available, the commitment could then be set at that level. Again, this approach limits possibilities but is certainly a good place to start for many organizations.

Either of these approaches can be used individually or in combination with each other. For organizations just starting the journey, they may choose to level the playing field by using the internal benchmarking approach. Organizations looking to get some quick wins and build momentum may choose to use the opportunity based approach. Organizations farther along the journey or looking to really disrupt things may choose the external benchmarking approach. Likewise, every organization should always be practicing external benchmarking as a sanity check.

A side note here: these benchmarking approaches can also be used to help with setting the financial objectives. Likewise, the information gained from setting the financial objectives can be used to set the KPI commitment levels.

6.1 Pillar goal development

Clearly, goals are a critical ingredient to strategic planning and a powerful tool in driving organizational performance. With a better understanding of goals and different approaches for determining the level at which goals should be set, it is time to leverage this knowledge and set the goals for each pillar. At this point, the focus is on setting goals for each of the pillar KPIs; goals related to the supporting metrics will be addressed later, so for now just focus on setting goals for each KPI. Continuing with the Smithville Foods example, let's work through this process.

Step 1: Choose goal-setting approach

The first step in setting a "smart" goal is to determine the approach that will be used to determine the level at which the goal will be set: External Benchmarking, Internal Benchmarking or Opportunity Based Benchmarking or another approach if that is preferred. Because each goal may require a different approach, the best way to work through this step is by addressing each KPI individually and choosing the best approach for each, as with the Smithville Foods example given here:

Table 6.1

KPI description	KPI number	Current KPI status	Goal-setting approach
Customer loyalty	Sales vs previous period (%)	Flat @ 0%	Internal benchmarking

Pillar: Customers

Customer loyalty is very specific to the organization; additionally, it would be very difficult to even find information from other organizations around this particular way of measuring customer loyalty. For this reason, external benchmarking would not be a good approach. Likewise, while opportunities can certainly be identified for improving customer loyalty, it would be difficult to calculate with any certainty the effect this would have on the loyalty score and thus opportunity based benchmarking would also not make a good approach. However, looking inwards at what other departments, divisions, or facilities have been able to achieve provides a good indication of what is possible and thus, in this situation, the internal benchmarking approach seems like the best way to go. See Table 6.1.

Table 6.2

KPI description	KPI number	Current KPI status	Goal-setting approach
Social impact	No. of meals provided	1,297	Internal benchmarking
Environmental impact	LCA score	17	Opportunity based benchmarking
Innovative products	% of sales from products 2 years or less	6.5%	Internal benchmarking
Quality products	No. of defective products to market/no. of products shipped	1.2%	Internal benchmarking

Pillar: Products

Under this pillar, like in the previous pillar, each of the KPIs are very organization specific (Table 6.2) and it may be hard to benchmark information from external organizations in these areas. So again, external benchmarking may not be the best approach for setting goals in these areas.

For the social impact KPI, opportunity based benchmarking could be used if the meals donated were a percentage of sales or something of that nature. The "opportunities" would be the level of sales or perhaps number of products sold. However, another way to go could be using internal benchmarking by looking at what was done in the past and setting a goal to top that going forward.

For the environmental impact KPI, any of the approaches may be used. Organizations may choose to start with internal and opportunity based benchmarking and once the low hanging fruit has been harvested, external benchmarking could be used to stretch the organization beyond their comfort zone.

For the innovative products and quality products KPIs, opportunity based benchmarking could pose some challenges due to the lack of comparisons available and therefore it may make more sense to just look internally and use the internal benchmarking approach.

Table 6.3

KPI description	KPI number	Current KPI status	Goal-setting approach
Efficiency	Units produced/ maximum output	78%	Internal benchmarking
Quality	No. of defects/units produced	2.1%	Internal benchmarking

Pillar: Operational excellence

Because both the quality and the efficiency KPIs are very specific to the organization and most organizations do not publicly report this information, using external benchmarking is probably not the best option here (Table 6.3). Opportunity based benchmarking is definitely an option since opportunities for improving both efficiency and quality can be identified and the impacts calculated rather accurately. Internal benchmarking is also a good way to approach this. Either one would be a good option in this case and perhaps both options could be used in combination or the organization could start with one and then move to the other over time.

Table 6.4

KPI description	KPI number	Current KPI status	Goal-setting approach
Employee safety	Total Recordable Injury Rate	18.3%	External benchmarking
Employee engagement	Gallup Q12 survey score	3.3%	External benchmarking

Pillar: People

For the people pillar (Table 6.4), both "employee safety" and "employee engagement" are standardized measurements that many organizations across many industries report. Additionally, "world class" benchmarks are readily available for these areas and thus, external benchmarking is a great approach in this situation. Of course, internal benchmarking could also be used along with opportunity based benchmarking.

Table 6.5

KPI description	KPI number	Current KPI status	Goal-setting approach
Response to business disruptions	Dollars lost due to business disruptions	$100,000	Internal benchmarking

Pillar: Resilience

In this situation (Table 6.5), the KPI "response to business disruptions" is extremely specific to the organization so external benchmarking would not really be an option at this point. Also, because the "opportunities" are relatively unknown, using the opportunity based benchmarking would not be the best solution either. This only leaves internal benchmarking as the approach for setting this goal.

At this point, the organization should have a clear picture of which approach will be used to set the commitment level for each KPI. Remember that when using Internal Benchmarking or the Opportunity Based Benchmarking, organizations may choose to also consider External Benchmarking as an additional comparison.

Before moving on to the next step it is important to note that this is only an example of how to choose the appropriate approach for setting the level of a goal or commitment. It is designed to provide a picture of the thought process required to choose the best approach. An organization may go through this very example on their own and decide to use different approaches than what has been presented. This is fine, the point is that organizations must consider different approaches, discuss the limiting factors to the various approaches and ultimately choose the approach that will work best for them.

Step 2: Setting a commitment level (goal)

With clarity around the approach to be used for setting the level of commitment for each of the KPIs, it is now time to actually set the level of commitment. Regardless of the approach chosen, this step is not a clear cut, scientific approach. It requires reflection, debate and an overall healthy level of discussion. Also remember, the end game is to set a "smart" goal, in other words a goal that is Specific, Measurable, Attainable, Realistic, Timely, and of course, Aligned. Remember that a key part of the alignment is ensuring that at a minimum, the cumulative money saved or generated from realizing the commitment levels must equal the financial objectives.

Because the goals are being set for each KPI under the different pillars, the "Specific" and "Measurable" pieces of the puzzle have already been addressed along with a portion of the alignment piece. This step is focused on making sure the goal is "Attainable" and "Realistic" using the approach chosen in the previous step while also ensuring realization of the goals will match the financial objectives to complete the alignment.

Important note here: a future step looks more specifically at a timeline for each goal but at this point the focus should be on setting a 1 year goal. In the next step, the organization will decide if there should be interim goals or milestones to support this one year goal. If a one year goal cannot be set at this point for one reason or another, leave it blank for now and address it in the next step. Let's continue with the Smithville Foods example to work through this.

Pillar: Customers

After reviewing the historic year over year sales results for key customers, it was determined that 5% improvements were routinely achieved (Table 6.6). Since the organization knows that this can be achieved, it is both realistic and attainable. Therefore, the organization has used this internal benchmark to set a 1 year goal of 5% sales

Table 6.6

KPI description	KPI number	Current KPI status	Goal-setting approach	1 year goal	Monetary impact
Customer loyalty	Sales vs previous period (%)	Flat @ 0%	Internal benchmarking	5%	$275,000

increase relative to the previous year. This is equal to $275,000 in additional revenue. Once this goal has been achieved, the organization can go back and identify if any higher growth rates have been periodically achieved and use this to set the bar a little higher for the following year. Eventually, this approach will reach its limit and the organization will have to use another approach to set a goal that goes beyond anything the organization has historically achieved.

Table 6.7

KPI description	KPI number	Current KPI status	Goal-setting approach	1 year goal	Monetary impact
Social impact	No. of meals provided as a result of money donated from sale of products	1,297	Internal benchmarking	1,400	Neutral
Environmental impact	LCA score	17	Opportunity based benchmarking	19	$60,000
Innovative products	% of sales from products 2 years or less	6.5%	Internal benchmarking	8%	$275,000
Quality products	No. of defective products to market/no. of products shipped	1.2%	Internal benchmarking	0.8%	$40,000

Pillar: Products

Under this pillar (Table 6.7), the KPIs of social impact, innovative products and quality products are all using the internal benchmarking approach. Like the example above, historical data was reviewed to determine what levels have been achieved historically to set the level of commitment for these KPIs. The environmental impact KPI is using opportunity based benchmarking. In this situation, the organization solicited ideas from various employees on how to improve the environmental impact of products. After reviewing each of the ideas, the organization identified the top ideas for implementation and determined the impact on the LCA score if each idea was implemented and set the commitment at that level. Since this commitment or goal is based on available opportunities, the organization knows that it is both realistic and attainable. Once again, the monetary impact was calculated for each of the commitment levels. In the case of the social impact KPI, they will need to find a way to provide more meals at a cheaper cost to ensure a neutral financial impact for this KPI.

Table 6.8

KPI description	KPI number	Current KPI status	Goal-setting approach	1 year goal	Monetary impact
Efficiency	Units produced/ maximum output	78%	Internal benchmarking	82%	$100,000
Quality	No. of defects/units produced	2.1%	Internal benchmarking	1.7%	$40,000

Table 6.9

KPI description	KPI number	Current KPI status	Goal-setting approach	1 year goal	Monetary impact
Employee safety	Total recordable injury rate	18.3%	External benchmarking	16.5%	$20,000
Employee engagement	Gallup Q12 survey score	3.3%	External benchmarking	3.8%	Neutral

Pillar: Operational excellence

For the operational excellence pillar (Table 6.8), the organization chose to use the internal benchmarking approach for both the efficiency and quality KPIs. By examining the overall historic data for the organization and historical performance at more granular levels, the organization was able to determine what level of performance was possible. This is a slightly different approach to just looking at the overall historical performance. By getting more granular in the analysis and examining what was achieved by different areas and/or processes within the organization, it highlights the performance that can be attained, even if it was only attained by a small slice of the organization.

Pillar: People

Using information gathered through research available online and through various public studies, the organization identified world class performance for both employee safety and employee engagement (Table 6.9). The world class level was too much of a stretch to achieve in 1 year so the organization dialed it back a little to what was determined to be more realistic for a 1 year time horizon. Once achieved, the level can be dialed up until the world class level has been achieved. For safety, the monetary impact can be calculated since it comes from reduced insurance premiums and the like. However, when it comes to engagement, it is difficult to predict the monetary impact and therefore Smithville chose not to record any positive financial impact for this commitment.

Revenue: $550,000	**Labor** $100,000	**Overhead** $165,000
Customer loyalty: $275,000	Efficiency $100,000	Environment $60,000
Innovative products $275,000		Quality product: $40,000
		Operations quality: $40,000
		Safety: $20,000
		Disruptions: $5,000

Table 6.10

KPI description	KPI number	Current KPI status	Goal-setting approach	1 year goal	Monetary impact
Response to business disruptions	Dollars lost due to business disruptions	$100,000	Internal benchmarking	$95,000	$5,000

Pillar: Resilience

Setting the level of commitment for this KPI is difficult for a few reasons. First, this is a relatively new type of focus for the organization so there is not much historical data available to review. Secondly, because it is a relatively new focus area, the organization does not have a lot of experience to go on. In this situation, Smithville Foods simply aimed to do better than last year. Once the organization has more historical performance data and gains more experience in this area, they will be to take a different approach to setting the level of commitment. This type of thing happens and organizations should not get too caught up in it. The lesson to be learned here is that when setting commitment levels, certain types of indicators may have some level of uncertainty and it is only over time that the organization will become more confident in setting commitment levels (Table 6.10).

At this point the organization should ensure that the monetary impact for the commitment levels align with the financial objectives that were set. For Smithville Foods, it looks like this:

It may take a few tries to get the cumulative monetary impact of the commitment levels to match up with the financial objectives but when matched, they complete an important piece of the alignment puzzle. Because these are the goals that will drive all improvement efforts, organizations can be assured that employees will be focused on things that bring the organization closer to its vision while also meeting financial objectives.

Another side note here: not all KPIs have to improve at once. In other words, it is not necessary to set higher commitment levels for each of the KPIs. Organizations may choose to keep certain KPIs flat. In fact, it can be beneficial to focus in on a few KPIs, especially in situations where organizations are short on resources.

In the next step, the organization will review each of the 1 year commitments to determine if there are interim milestones that need to be set to support this 1 year goal.

Step 3: Setting milestones

With clear 1 year commitments established for each KPI, it is important to determine if interim goals or milestones makes sense. While it is valuable to set 1 year or longer term goals, shorter term goals should also be set where possible to support the longer term goals. If the goal reaches out too far, people tend to lose focus or procrastinate. Setting interim goals or milestones will help keep people focused, enable stronger accountability, and better facilitate management of the overall commitment.

When it comes to setting milestones, the knee-jerk reaction is to automatically set them all at the same time intervals. While this may work, it is suggested that each commitment be addressed individually and a specific timeline chosen that best suits the situation. For example, depending on the goal, availability of data, etc., it may make sense to set shorter timelines for some goals such as monthly or even weekly, whereas for others it may make more sense to stretch out the time line to quarterly or even semi-annual.

Table 6.11

KPI description	KPI number	Current KPI status	Goal-setting approach	1 year goal	Milestones
Customer loyalty	Sales vs previous year (%)	Flat @ 0%	Internal benchmarking	5%	Semi Annual

Pillar: Customers

Customer loyalty is something that takes time to move the needle on and sales may be driven by seasons or cycles. In this case semi-annual milestones make sense, anything more frequent may provide a skewed picture of actual performance (Table 6.11).

Table 6.12

KPI description	KPI number	Current KPI status	Goal-setting approach	1 year goal	Milestones
Social impact	No. of meals provided	1,297	Internal benchmarking	1,400	Quarterly Q1: 350, Q2: 700 Q3: 1050, Q4: 1400
Environmental impact	LCA score	17	Opportunity based benchmarking	19	1 Year
Innovative products	% of sales from products <2 years	6.5%	Internal benchmarking	8%	1 Year
Quality products	No. of defective products to market/no. of products shipped	1.2%	Internal benchmarking	0.8%	Quarterly Q1: 1.1% Q2: 1.0% Q3: 0.9% Q4: 0.8%

Pillar: Products

For the social impact and quality products commitments, data is available, and performance changes regularly and can be easily tracked (Table 6.12). For these reasons, it makes sense to support the

1 year goal with more frequent, quarterly milestones. With frequent milestones, the organization must set the goal level for these milestones as noted above.

For both the environmental impact and innovative products commitments, a 1 year goal makes the most sense. Both of these areas require substantial time to improve and performance may be hard to measure frequently. However, while setting "hard" milestones, may not make sense, setting "soft" milestones could make sense. For example, setting quarterly milestones to have certain products launched for the innovative products goal or setting a milestone to have environmental impact reductions identified will help to keep things moving and keep people focused.

Table 6.13

KPI description	KPI number	Current KPI status	Goal-setting approach	1 year goal	Milestones
Efficiency	Units produced/ maximum output	78%	Internal benchmarking	82%	Quarterly Q1: 79%, Q2: 80% Q3: 81%, Q4: 82%
Quality	No. of defects/ units produced	2.1%	Internal benchmarking	1.7%	Quarterly Q1: 2.0%, Q2: 1.9% Q3: 1.8%, Q4: 1.7%

Pillar: Operational excellence

Operational performance can change quite rapidly and reliable data is readily available so it makes sense here to set shorter milestones to support the 1 year goal (Table 6.13). Some organizations may even choose to set monthly or even weekly milestones.

Table 6.14

KPI description	KPI number	Current KPI status	Goal-setting approach	1 year goal	Milestones
Employee safety	Total recordable injury rate	18.3%	External benchmarking	16.5%	Monthly Improve by 0.15% month over month.
Employee engagement	Gallup Q12 survey score	3.3%	External benchmarking	3.8%	Biannually 6 Month: 3.5%

Pillar: People

Safety performance, like operational performance, is very dynamic and data is readily available so it makes sense to set shorter milestones such as monthly or even weekly in this area. Employee engagement on the other hand, takes a little longer to enact change and measuring performance is more involved since it requires employees to complete a survey and thus a bi-annual milestone would make the most sense here. More frequent milestones can be set in this area but employees may experience "survey fatigue" leading to unreliable performance data among other challenges (Table 6.14).

Table 6.15

KPI description	KPI number	Current KPI status	Goal-setting approach	1 year goal	Milestones
Response to business disruptions	Dollars lost due to business disruptions	$100,000	Internal benchmarking		Annual

Pillar: Resilience

Since this goal will only change when there is a business disruption, it does not make sense to look at milestones but rather just review annually or when a disruption occurs (Table 6.15).

At this stage, organizations should have a good handle on the level of commitment to be achieved over the next year and set any interim milestones to gauge and manage performance towards the 1 year goal.

Step 4: Assigning ownership

With "smart" goals aka "commitments" in place and milestones identified for each, someone must ultimately be responsible for achieving them. Assigning ownership to each commitment is critical to actually achieving those commitments, which is a crucial piece to the overall success of the strategic plan. If the goals or commitments are not achieved, performance under each pillar does not improve, the pillar missions are not realized, the vision cannot be achieved, and ultimately the purpose cannot be fulfilled.

When assigning ownership, it is suggested that it be assigned to one individual. Also, resist the temptation to automatically assign ownership to an individual for all the commitments that fall under a

Table 6.16

KPI	Current	1 year goal	Milestones	Ownership
Customer loyalty	0%	5%	Annual	President
Social impact	1297	1400	Q1, Q2, Q3, Q4	Director, sustainability
Environmental impact	17	19	Annual	Director, sustainability
Innovative products	6.5%	8%	Annual	Vice President sales & marketing
Quality products	1.2%	0.8%	Quarterly	Chief Operating Officer
Efficiency	78%	82%	Quarterly	Vice President operations
Quality	2.1%	1.7%	Quarterly	Engineering director
Employee safety	18.3%	16.5%	Monthly	Environmental Health and Safety director
Employee engagement	3.3%	3.8%	Biannual	Director, Human Resources
Response to disruptions	$100,000	$95,000	Annual	Chief Operating Officer

particular pillar. For example, try to avoid assigning all the operational excellence commitments to the operations leader. While this may ultimately make the most sense, work through each of the goals individually to assign ownership. Since these goals are high level goals related to the KPIs, the ownership should also be at a higher level in the organization such as Executives or Senior Managers. In the next step of the strategic planning process, these commitments will be cascaded down and at that time, ownership will be assigned at different levels across the organization but at this point, the focus is on the higher level goals or commitments and thus, higher level ownership (Table 6.16).

For Smithville Foods, there are situations where one individual is responsible for more than one KPI; however, the ownership is relatively spread out amongst the senior leadership team.

At this point, there are smart goals aka commitments in place and ownership for each commitment has been assigned. This is a good time to go back and update the KPI dashboard with the commitment levels for each KPI as shown in the Smithville Foods example below.

Notice in the example, that there is still one column blank. This is fine; in the next chapter, the variance will be determined and it can be entered into the dashboard at that time – so leave it blank for now.

KPI dashboard					
Pillar	**KPI description**	**KPI number**	**Current**	**Goal**	**Variance**
Customers	Customer loyalty	Sales vs previous period (%)	0%	5%	
Products	Social impact	Meals provided	1297	1400	
	Environmental impact	Lca score	17	19	
	Innovative products	% of sales < 2 years	6.5%	8%	
	Quality products	Defects / products shipped	1.2%	0.8%	
Operational excellence	Efficiency	Units Produced / max output	78%	82%	
	Quality	No. of defects / units produced	2.1%	1.7%	
People	Employee safety	Trir Rate	18.3%	16.5%	
	Employee engagement	Q 12 Score	3.3%	3.8%	
Resilience	Response to business disruptions	Dollars lost to business disruptions	$100,000	$95,000	

The foundation has now been laid to complete the final step of the strategic planning process which is to develop the programs that will be put in place to achieve these commitments.

Progress indicator

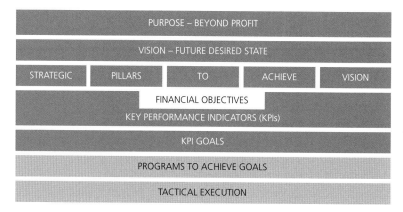

Fig. 6.1 The sixth step in the strategic planning process is complete – commitment levels have been established for each KPI and a clear owner assigned to each.

CHAPTER 7

Adding Value – Program Development

Doing the right thing is more important than doing the thing right.

Peter Drucker

Twenty minutes a day equates to 10 days a year. Multiply this by the number of employees in an organization and the figure becomes staggering. Even in small organizations with, say, 50 employees, if each employee spent 20 minutes a day doing something, at the end of the year it would equate to 500 days. Now imagine the 20 minutes was spent working on the wrong stuff – stuff that does not add value or is not required. At the end of the year, even a small organization would have wasted 500 days of valuable time.

Competing and winning in today's fiercely competitive marketplace requires that *all* employees deliver value *all* the time. Any time spent on the "wrong stuff" is wasted time that depletes value generated by the organization and shifts advantage to a competitor. Even what seems like a small amount of time can add up over a year or a quarter or even a month. This does not mean employees should be driven into the ground like slaves or monitored by "big brother." Employees are going to goof around, come in late, leave early, check out Facebook or surf the web. So what? This is normal; in fact, some organizations

---

Purposely Profitable: Embedding Sustainability into the DNA of Food Processing and Other Businesses, First Edition. Brett Wills.

even encourage a more laid-back office with no set office hours, areas for employees to socialize, work out, and do many other "non-work" related things. In fact, many of these organizations with a "non-traditional" work environment show a much higher degree of engagement, productivity, and overall performance. Why is this? Well, the key is that when employees *are* working, they work hard, are laser focused, and most importantly, they are working on the right stuff.

The trick is figuring out what the "right stuff" is. The high level answer is simple: the right stuff to be working on are the things that will bring the organization closer to its vision. This sounds great in theory but is not very pragmatic as it does not specify the "things" or "stuff" that should be the focus of employee time and effort.

In other words, it does not identify the specific programs and/or initiatives that should be put in place to drive performance and bring the organization closer to its vision. The focus of this step in the strategic planning process is to do exactly that – identify the specific programs and/or initiatives to be put in place. The kicker is that through completing the previous steps in the process, the foundation has been laid to identify the programs and/or initiatives that will ensure employees are working on the right stuff. Why is this? Let's take a closer look at the overall process again as depicted in Fig 7.1 for the answer.

The programs and initiatives put in place will support achievement of the goals, when the goals are achieved, the KPI performance improves. Since the KPIs measure performance towards achieving the pillar missions, which enables the organization to achieve its vision, the organization knows it is working on the right stuff because

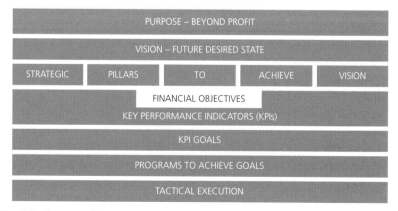

Fig. 7.1 The overall process

the programs are aligned directly to the vision. The other great thing here is that if it is not the right stuff, the KPIs will indicate this and modifications can be made to the programs.

Now, just to clarify something here, the programs being referred to are the "added" things, the improvement projects that will be worked on to improve current performance. It is not necessarily the regular daily activities that must take place in order to keep the organization ticking, although the improvement and daily activities are certainly connected. Things like entering an order or building a widget or dealing with a customer concern are not the focus here. A more efficient way to enter an order or a better way to build a widget or a quicker way to address customer concerns is the focus here. Over time, these improvement activities will drive changes in the day to day activities and processes to become the new standard for how things are done.

Having said that, the KPIs developed earlier on in the process will provide a measure of whether the day-to-day activities are performing as desired. If they are not, it will show up through the KPIs and improvement programs can be implemented to correct performance.

The programs could be something like redesigning a product to improve environmental performance or implementing a continuous improvement program such as Lean or Six Sigma to improve operational efficiency.

Choosing the right improvement programs to put in place can be tricky. Organizations must understand the different types of programs available, the problems they solve, and the value they offer, then choose the best one(s) to achieve the goals. Too many organizations fall down in this area and end up choosing the wrong programs causing employees to work on the wrong stuff and then fail to realize the desired results. The rub is that there are so many programs available and they come with an infinite combination of possibilities, it makes it very challenging to choose the right ones.

The "Tools and resources" section at the end of this book provides an overview of the main tools, resources, processes, etc., that organizations can use to identify/develop/inspire the programs that will be put in place to achieve the goals that have been set. The tools and resources vary from standard, well-known processes such as Lean or Six Sigma and the various tools under these such as 5S, Value Stream Mapping or the "Fish Bone" problem solving tool to sustainability related tools such as Carbon Footprinting, Life Cycle Analysis, or Biomimicry. While it is outside the focus of this book to provide a great amount of detail on each, a sufficient overview is provided to enable organizations to determine whether a specific tool may be a good fit and worth further investigation.

7.1 Program development

Similar to previous steps in the Strategic Planning process, this part of the process is not prescriptive per se. Rather, it provides the steps required and the type of thinking needed to identify the proper programs that will ensure people are working on the right stuff to achieve the goals or commitments that have been set for each KPI. Understanding what tools and resources are available is a good start but not enough to get the right programs in place to drive performance, so let's continue with the Smithville Foods example and work through the steps for putting the right programs in place.

Step 1: Gap analysis

The first step in putting the right programs in place is to understand the gap between where the organization is today relative to where it wants to be in the future (the Goal – aka Commitment). This is a rather simple step if the previous parts of the strategic planning process have been completed. By identifying the KPIs and building the baselines for them, the organization will have a clear picture of where it is today. By setting clear commitment levels for each of these KPIs, the organization will have a clear picture of where it wants to be in the future. The difference between the current state and the future desired state is the "gap."

Organizations must perform a high-level gap analysis and identify, for each KPI, the gap between where they are now and where they want to be. Using Smithville Foods as an example, it could look something like Table 7.1:

Clearly this is a simplified example but the concept should be clear; identify the gap between the organization's current performance and where the organization would like its performance to be 1 year in the future. This is a fundamental step required to complete the next step, which is critical for identifying the right programs to put in place because it clearly identifies the extent of improvement that is required. Also, this point is a good time to update the KPI dashboard with the final piece of missing information as shown in the Smithville KPI Dashboard example.

Table 7.1

KPI	Current	1 year goal	Gap
Customer loyalty	0%	5%	5%
Social impact	1297	1400	103
Environmental impact	17	19	2
Innovative products	6.5%	8%	1.5%
Quality products	1.2%	0.8%	0.4%
Efficiency	78%	82%	4%
Quality	2.1%	1.7%	0.4%
Employee safety	18.3%	16.5%	1.8%
Employee engagement	3.3%	3.8%	0.5%
Response to disruptions	$100,000	$95,000	$5,000

KPI dashboard					
Pillar	**KPI description**	**KPI number**	**Current**	**Goal**	**Variance**
Customers	Customer loyalty	Sales vs previous period (%)	0%	5%	**5%**
Products	Social impact	Meals provided	1297	1400	**103**
	Environmental impact	Lca score	17	19	**2**
	Innovative products	% of sales < 2 years	6.5%	8%	**1.5%**
	Quality products	Defects / products shipped	1.2%	0.8%	**0.4%**
Operational excellence	Efficiency	Units produced / max output	78%	82%	**4%**
	Quality	No. of defects / units produced	2.1%	1.7%	**0.4%**
People	Employee safety	Trir rate	18.3%	16.5%	**1.8%**
	Employee engagement	Q 12 score	3.3%	3.8%	**0.5%**
Resilience	Response to business disruptions	Dollars lost to business disruptions	$100,000	$95,000	**$5,000**

Step 2: Pareto analysis

With a clear understanding of the gap between the current state and the future desired state, the next step in the process is to determine why the gap exists. The focus here should not be on simply identifying the surface-level symptoms but rather trying to get down to the root cause(s) of the problem. Unfortunately, this can be challenging and it is not always possible to pinpoint the exact root cause(s) immediately and, furthermore, there could be several underlying factors contributing to the gap. Luckily there are a number of tools and resources available to help. Refer to the "Problem solving" section of the "Tools and resources" chapter to learn more about the different tools and resources that can provide support in completing this step. After completing this step, the organization should have identified the root causes for the gap which will look something like the Smithville example shown in Table 7.2.

Table 7.2

KPI	Current	1 year goal	Gap	Root cause(s)
Customer loyalty	0%	5%	5%	Price, quality
Social impact	1297	1400	103	Cost of meals
Environmental impact	17	19	2	Packaging material, manufacturing process
Innovative products	6.5%	8%	1.5%	Budget, formal program
Quality products	1.2%	0.8%	0.4%	Damages during shipment
Efficiency	78%	82%	4%	Changeovers, poor housekeeping
Quality	2.1%	1.7%	0.4%	Poor housekeeping, equipment malfunction due to lack of maintenance
Employee safety	18.3%	16.5%	1.8%	Poor housekeeping
Employee engagement	3.3%	3.8%	0.5%	Lack of performance reviews
Response to disruptions	$100,000	$95,000	$5,000	Lack of decentralized systems

Once again, a simple example but the point should be clear; identify and document the reasons for the gap. This may require the use of one or more of the tools covered in the Tools and Resources section or it may involve simply pulling in those folks who are closest to the problem or issue to gain better insight on the issue. This could be a senior level person but many times it is the front line folks who are closest to the issue and have the greatest insight as to why certain things are happening the way they are. Many times these folks also have solutions identified but because no one has ever asked them, the solutions remain hidden.

It is amazing how surprised organizations get when they ask a front line employee about a particular problem and they not only know why the problem is occurring but they also have a solution to fix it.

There are many elaborate tools and resources out there and they do have their place but often, the answer is as simple as talking with the folks who spend day in and day out looking at the problem.

Step 3: Program identification

With the reasons for the Gaps identified, now it is time to begin closing those gaps by identifying the right solution (aka program) to put in place. The program may be centered around a specific tool such as 5S or an Energy Audit or possibly involve a broader approach such as Lean or Six Sigma. It may involve a softer approach such as the forming of a focused team or completion of training. Whatever the program may be, it is critical that it be the right tool or solution to solve the problem(s) identified in the previous step.

It is disappointing how many organizations skip directly to this step of program identification without completing any of the background work required to get here and correctly identify the right program to put in place. Too many organizations choose to use a specific tool or resource and then create a program around it simply because it happens to be the current fad or the latest focus in the management or consulting world and not because it is the right solution for their particular situation. The organization must work through each of the root causes to determine the best program to put in place that will address the cause and realize the goal. Continuing with the Smithville example, this step could look something like the Table 7.3.

Table 7.3

KPI	Gap	Root cause(s)	Programs
Customer loyalty	5%	Price, quality	Supplier negotiation program, product re-design based on biomimicry Program
Social impact	103	Cost of meals	Meal cost reduction program
Environmental impact	2	Packaging material, manufacturing process	Package lightweighting program, lean and green program
Innovative products	1.5%	Budget, formal program	Establish RnD budget, build RnD strategy
Quality products	0.4%	Damages during shipment	Logistics improvement program
Efficiency	4%	Changeovers, poor housekeeping	SMED program, 5S program
Quality	0.4%	Poor housekeeping, equipment malfunction due to lack of maintenance	5S program, TPM program
Employee safety	1.8%	Poor housekeeping	5S program
Employee engagement	0.5%	Lack of performance reviews	Implement performance management system
Response to disruptions	$5,000	Lack of decentralized systems	Design decentralized data system

Step 4: Program metric development

At this point, the programs (aka solutions) to achieve each of the goals should be identified. The next step is to set up a means for measuring performance of these programs. Since the measurement here is for the programs that support the goals or commitments for the KPIs, these measures are considered metrics instead of KPIs. These metrics will be a combination of leading indicators to measure the input that will drive performance combined with lagging indicators to measure performance output or results.

To avoid redundancy, this chapter does not provide the level of detail for developing measures that was provided in the KPI chapter. However, even though the focus here is on metrics vs KPIs, the process remains the same. The end game is to identify a means to measure performance, in this case, performance of the programs that are put in place to achieve the commitment levels for each KPI that were set earlier in the strategic planning process. The only difference here is that organizations are encouraged to supplement the metrics that will measure performance (lagging indicators) with measures that will drive performance (leading indicators). To clarify this point let's continue with the Smithville Foods example (Table 7.4).

In the example, note the program for improving Customer Loyalty. As a reminder, this program is designed to reduce costs through supplier negotiations. By doing this, it will address the issue identified through the customer loyalty survey that customers would be more loyal if the product was offered at a more affordable price. In this case, the natural metric to measure performance would be the amount of dollars saved through negotiations – a lagging metric. This is great and required but remember back to Chapter 5 on KPIs – a great way to drive this lagging indicator is to support it with a leading indicator – like measuring calories burned and caloric intake supports the lagging indicator of weight lost for a diet program. In this case a leading indicator could be the number of supplier negotiations completed. The more negotiations that are completed, the more dollars that will be saved (if not, then maybe some training needs to be completed to improve negotiation skills).

There may be additional or different leading indicators that could be put in place here but the point is to support the lagging metric with a leading metric to actively drive performance. At this point in the strategic planning process, the focus is on building ways or mechanisms to drive performance actively and the leading metric is a

Table 7.4

KPI	Programs	Leading metric	Lagging metric
Customer loyalty	Supplier negotiation program	No. of negotiations completed	Dollars saved through negotiations
Social impact	Meal cost reduction program	No. of cost reduction initiatives implemented	Dollars cost savings
Environmental impact	Package lightweighting program	No. of lightweighting initiatives implemented	Amount of weight removed from packaging
Innovative products	Establish RnD budget	N/A	RnD budget established
Quality products	Logistics improvement program	No. of corrective actions implemented to reduce damages during shipment	No. of damages due to shipping
Efficiency	SMED program	No. of changeover reductions implemented	Total changeover reduction time
Quality	TPM program	TPM program implemented	No. of defects due to equipment malfunction
Employee safety	5s program	5S audit score	No. of injuries due to poor housekeeping
Employee engagement	Implement performance management system	Performance management system implemented	% of performance reviews completed
Response to disruptions	Decentralized system	Decentralized system design completed	% of organization on decentralized system

proven way to do this. This same thinking was used to develop the remainder of the leading and lagging metrics for the other programs in the example provided. Using this approach and thinking, organizations can work to develop their own set of leading and lagging metrics for the programs that were identified.

Step 5: Program goal development

Like with the KPIs, the next step here is to set a goal or commitment for each metric. Again, to avoid redundancy, there will not be the level of detail as provided in Chapter 6 on goals since the process is

Table 7.5

KPI	Leading metric	Leading metric goal	Lagging metric	Lagging metric goal
Customer loyalty	No. of price negotiations completed	4	Dollars saved through negotiations	$25,000
Social impact	No. of cost reduction initiatives implemented	5	Dollars cost savings	$3 per meal
Environmental impact	No. of lightweighting initiatives implemented	3	Amount of weight removed from packaging	1 lb
Innovative products	N/A	N/A	RnD budget established	Establish budget
Quality products	No. of corrective actions implemented to reduce damage during shipments	1 per month	No. of damages due to shipping	Reduce by 50% vs previous year
Efficiency	No. of changeover reductions implemented	7	Total changeover reduction time	25% reduction
Quality	TPM program implemented	Implement within 6 Months	No. of defects due to equipment malfunction	Reduce by 50%
Employee safety	5S audit score		No. of injuries due to poor housekeeping	0
Employee engagement	Performance management system implemented	Implemented	% of performance reviews completed	80%
Response to disruptions	Decentralized data systems designed	6 months	% of organization on decentralized system	50% within 1 year

the same. Continuing with the Smithville Foods example, the end result of this step can look something like Table 7.5:

Some of the goals in this example are softer goals in that they do not set a hard reduction or improvement number but rather indicate that a task must be completed such as implementing a Performance Management System. Like when setting goals or commitment levels for KPIs, the goals for each of these metrics should also be supported with a completion date and assigned an owner who will be ultimately responsible for achieving the goal.

Step 6: Action plan development

The final step in program development is where the rubber actually hits the road and people actually start getting stuff done. It's great to have goals or commitment levels set but they are not achieved by walking into work one day and doing one big thing, goals are achieved when a whole bunch of small little actions are completed that result in the culmination of a noticeable change. For example let's take the above goal for the program designed to improve environmental impact through a one pound lightweighting of packaging. It's not like all of a sudden someone walks into work one day and boom there is a pound of weight gone from the packaging. There are a bunch of small actions that are required to get this weight off. Potential consultants must be identified, quotes and proposals must be received, a consultant must be chosen, audit of the packaging scheduled, a report produced with opportunities for lightweighting identified, opportunities prioritized, the new design prototyped, testing completed, certifications put in place, and so on. None of these actions are big things per se, but by completing all these little things, it culminates in one big thing getting done or one goal being achieved. These groups of small little actions required to achieve the program goal can be called tactical action plans. These tactical action plans will be what drives getting things done and what ultimately enables organizations to achieve the goals and commitments that have been set.

There are certain components that should be included in the tactical action plan to maximize chances of success. Each action plan should contain the following five components:

1 Name of the action item
2 Description of the action item
3 Due date for the action item
4 Owner of the action item
5 Current status of the action

The tools and resources section found at the end of this book contains an example template that can be used to quickly develop a form for completing the action plans. It is strongly suggested that these action plans be developed by the people or teams that will be executing them. By having the person or team that is responsible for executing the action plan, also developing the action plan, it creates another layer of ownership and engages these folks in the initiative early on.

While it may seem simple or basic, many organizations are dismissing simple basic management concepts such as these tactical action plans and finding out, that while very simple and basic concepts, they are very powerful and necessary to drive performance forward. Do not underestimate the power of these action plans – they are critical to driving performance and lay out, step-by-step, the actions required to achieve the desired performance. The next chapter will look at how to actively manage these action plans to ensure they get executed.

Having followed each of these steps in this part of the strategic planning process, the organization should now have a clear set of programs supported by tactical action plans that will drive day-to-day improvement activities which are aligned to achieving the goals for each KPI. Achieving the goals for each KPI will improve that KPI which will move the organization closer to realizing the mission for each strategic pillar which will drive achievement of the vision to ultimately fulfill the purpose. In other words, the organization has developed a Strategic Plan that has culminated in a number of small action items that are completely aligned all the way back up to the achieving the Vision and fulfilling the Purpose. Because of this alignment, the organization can be confident that employees are working on the right stuff! Development of the programs and supporting action plans is one of the last major steps in the strategic planning process for developing a strategy that is sustainable. If organizations have followed the process to this point, then a clearly laid out strategic plan that integrates all the dimensions of sustainability and is fully aligned will be developed. This means the organization now has a strategy that is sustainable. Congratulations! This is definitely something that should be celebrated and all the people involved in the process should be recognized. Not so fast though, the plan is not completely finished at this point. In fact, there is one more step in the process that some argue is one of the most important pieces of the puzzle. It does not have to do with developing the strategy but rather it specifically deals with *executing* the strategy.

Progress indicator

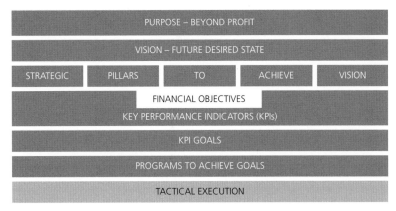

Fig. 7.2 The seventh step in the strategic planning process is complete – programs to achieve the commitments for each KPI have been identified and a tactical action plan developed for each Program with a clear owner assigned to each.

CHAPTER 8

Getting Tactical – Strategy Execution

Strategy gets you into the game. Execution IS the game.

Sam Geist

Developing a strategy is one thing, executing it is another. A strategy in itself does not build success, effective execution of the strategy does. Successful execution of the strategic plan is what separates the great organizations from the good organizations. Organizations that place emphasis on execution are able to produce extraordinary results.

Unlike the process for developing a strategic plan, there is no clear-cut process for developing a plan to execute the strategy. However, there are certain characteristics and conditions shared by organizations that are successful with strategy execution. What follows is an in-depth look at these different characteristics and conditions. Garnered from countless days spent inside organizations studying why some are successful with execution and others are not, these practices are proven, practical, and pragmatic.

The interesting thing here is that there is no "secret sauce" to execution that is taught behind closed doors at some Ivy League graduate school or some classified management technique privy to only high powered executives. In fact, by studying the practices of successful

Purposely Profitable: Embedding Sustainability into the DNA of Food Processing and Other Businesses, First Edition. Brett Wills.
© 2016 John Wiley & Sons, Ltd. Published 2016 by John Wiley & Sons, Ltd.

organizations and their leaders when it comes to execution, it surprisingly revealed that the characteristics and conditions required for successful execution are fairly simple. What separates the great organizations from the good organizations in terms of strategy execution is that the great organizations have the basics down pat and relentlessly practice them. The mediocre organizations and leaders dismiss the basics and fail to achieve the desired results. Let's look at what the great organizations are doing to successfully execute their strategy.

8.1 Communicating the strategic plan

Successful execution of the strategy requires that the entire organization be aligned, that all the cogs in the wheel be moving in the same direction. In order for this to happen, every single person in the organization must be crystal clear on where the organization is going and – at a high level – how it is going to get there. Additionally, expectations for performance and behaviors must also be clear. In other words, every person in the organization must understand and buy in to the strategic plan. It may seem simple and basic but the only way people are going to understand and buy into the strategic plan is if it is effectively communicated to them. The great organizations and leaders do this very well and they do it relentlessly. They never stop communicating the plan, never stop reinforcing the vision, or reinforcing the factors critical to success. The mediocre organizations think they do this well but in reality, they do not.

Communicating the strategic plan is about more than simply throwing a visual up on a wall or running through a presentation on the purpose, vision, pillars, KPIs, goals, etc. Organizations who are successful at executing their strategy go far beyond simply "communicating" the strategy and spend a considerable amount of time explaining the strategy to all staff, ensuring each person clearly understands it and their role in bringing it to life. This cannot be stressed enough and while it seems fairly simple and rudimentary, many organizations fail with this and therefore fail in successfully executing the strategy.

While it is ultimately the responsibility of the organization's leader, such as the President or CEO, to effectively communicate the strategy, the responsibility does not end there. The executive team also shares in this responsibility as do the other senior leaders and managers.

Shared responsibility for communicating the strategy becomes increasingly important the larger the organization. Each of these folks is responsible for ensuring people in their department or area clearly understand the strategy and their role in bringing it to life.

Not only is it critical that every senior leader understand their responsibility for communicating the vision, the organization should have a standardized method of communication. Each senior leader should be using the same or similar approaches for communicating the strategy, which will ensure that everyone across the organization is receiving the same message and all people are aligned to the same picture of the future.

Now, effectively communicating the strategic plan is a lot more art than it is science but there are certain traits or qualities that lead to effectively communicating the strategy. Looking at each piece of the strategic plan in turn and exploring how to effectively communicate that piece of the puzzle will help to more effectively communicate the overall strategic plan. Before doing that, however, it is important for organizations to share the model that was used to develop the strategic plan and the different steps of the process that were taken to develop the strategy. This should start with an overview of Sustainability to ensure that all folks have a solid understanding of Sustainability and how the strategic model integrates each dimension into the overall strategy through the pillars.

By providing this high-level overview, folks will better understand each piece as it is being communicated and help them better digest the overall strategy. Let's now look at ways to effectively communicate each piece of the strategic plan.

8.1.1 Purpose

Remember that the Purpose is why the organization exists, the reason for which it opens the doors every day, and what everyone is ultimately working to fulfill – this must be communicated to everyone. Furthermore, effectively communicating the Purpose will mean that employees become emotionally attached to it. This is a tall order but here are a few tips to help:

1 **Be authentic**: Organizations must be genuine when talking about their purpose because people can immediately see right through a fake exterior. Lack of authenticity will prevent folks from growing

any sort of emotional attachment and even worse, cause people to become resentful. The thing is, authenticity cannot be faked which means the leader and other senior folks in the organization must truly believe in the purpose. If these folks do not believe in the purpose then either it is not the correct purpose or they are not the right people and something must be changed.

2 **Be passionate**: When talking about the Purpose leaders must show passion for it. Talking about the purpose in a soft, monotone, lazy voice is not going to convey passion for the purpose and get people fired up about it. Fortunately, if the leaders really believe in the Purpose, then being passionate about it should come through naturally.

3 **Make it tangible**: In order for employees to become emotionally attached to the purpose they need something to relate to, something more than words on a piece of paper. Leaders need to paint a clear picture of what the future will look like if the Purpose is fulfilled. Using the Smithville Foods example, what will the future look like if health and well-being are improved? Will this mean that childhood obesity will be a thing of the past? Will this mean that childhood diabetes will only be talked about in history books? Will this mean that a much smaller amount of the population will be stricken with certain illnesses? Whatever the answers may be, the organization must answer them and paint a clear picture of what the future will look like when the Purpose is fulfilled.

Clearly this is not the be all and end all for effectively communicating the Purpose but rather some things to think about and help get started with effectively communicating the Purpose.

Vision

Discussed in greater detail in Chapter 2 on Vision Development, here is a reminder of some ways to effectively communicate and build a shared vision:

1 **Making it tangible**: Like with the Purpose, the Vision must be brought to life by painting a clear picture of what the organization will look like when it achieves the Vision.

2 **Aligning daily activities to the vision**: Every single employee must understand how their daily activities contribute to achieving the Vision and how failure to properly complete their duties can

prevent the vision from being realized. One way to do this is by using the analogy of a bank account. Achieving the Vision requires a certain amount of deposits into the vision bank account. Every time an employee does their job correctly it is like making a small deposit in the account, which over time builds up to realize the vision. However, every time an employee does not do their job correctly it is like making a small withdrawal from the bank account. A few small withdrawals here and there are tolerable, but make too many withdrawals and the company will never build up the savings needed to achieve the vision. On the extreme side, if there are excessive withdrawals from the account there will be no savings and the organization will go bankrupt – both figuratively and literally.

In addition to creating a shared vision, the organization must clearly link the Vision to achieving the overall Purpose by showing how achieving the vision will, over time, help to fulfill the purpose. For example, with the Smithville Foods example, the vision of being: *"North America's preferred provider of fresh produce solutions"* will allow them to provide healthy and fresh food for current and future generations and therefore fulfill the Purpose *"To improve the health and well-being of current and future generations."*

8.1.2 Pillars

While it can be a little time consuming, it is critical that organizations take time to clearly communicate each of the pillars to all employees. If the organization took the time at the start to clearly communicate the model that was used in developing the strategic plan, some of the hard work has already been completed as employees will already understand the concept of the pillars and how they are the focus areas critical to success, achieving the vision, and ultimately fulfilling the purpose. Assuming the model for developing the strategic plan has already been communicated, let's look at some key drivers for effectively communicating the pillars:

1 **Explain the pillar missions**: Like with the purpose and vision, the organization must paint a clear picture of what is meant by the pillar mission statement. For example, with the Smithville Foods example, the "Customer" pillar has a mission to: *"Build loyal partnerships with our customers through a focus on radical transparency and nimble customer service."* Employees must understand what is meant

by a loyal partnership, what is meant by radical transparency and nimble customer service. Does this mean that customers never buy from another organization, that all dirty laundry will be aired with every customer, that every customer will receive service within 30 seconds? Whatever the answer may be, the organization must clearly communicate the meaning of the mission to each employee.

2 **Link to Sustainability**: In some situations the link to each dimension of Sustainability may not be explicit and thus requires further explanation. The organization must ensure that people understand how each dimension of Sustainability has been covered across the pillars. This can be done when explaining the mission or as a separate exercise after each mission has been explained.

3 **Link to daily activities**: Just like with the vision, each employee must understand how their daily activities impact each pillar. Illustrating how daily activities impact each pillar will further solidify how these daily activities impact the vision and better help them make the connection between what seems like a mundane task and how completing the task contributes to a greater purpose.

8.1.3 KPIs

A fairly boring and dry subject to talk about, nevertheless it is important. There is no real magic here – when communicating KPIs employees should understand:

1 KPIs measure performance under each pillar
2 What the KPI number means
3 Why each KPI was chosen
4 What the current KPI is
5 How their daily activities will affect this KPI
6 How it will be used as part of the performance reviews

8.1.4 Goals (aka Commitments)

Again, when it comes to communicating the goals or commitment levels, it is a relatively straightforward and quick exercise. Employees need to understand:

1 A commitment level has been set for each KPI
2 How that commitment level was determined
3 Which commitments apply to them
4 How people will be held accountable to the commitments

8.1.5 Programs

If all the other pieces of the strategic plan have been effectively communicated, explaining the last piece of the puzzle should be relatively straightforward. Employees need to understand a few things about the programs. First, they must understand how the programs are improvement initiatives to support achievement of the goals for each KPI. Secondly, employees should understand that these programs will be driven by detailed action plans that are developed by them. Finally, each employee must understand their role in the program(s).

If all the above steps have been completed, each employee should have a clear understanding of the strategic plan and what the future will look like by executing this plan. Again, do not underestimate the power of this basic concept, take the time to properly communicate the strategy.

8.2 Cascading the strategic plan

To facilitate alignment of the strategic plan across the entire organization, it may need to be cascaded down to the different divisions, facilities, departments, etc. This can get a little tricky in some situations, so let's explore a straightforward approach for doing this effectively. The easiest way to understand how to do this is to start by looking at a simple scenario of a smaller less complex organization and work up towards a larger more complex organization.

In the scenario of a small to medium-sized organization where there is only one facility with one division, there is no real cascading required. These organizations can follow the steps for effectively communicating the strategy and ensure that each department understands which KPIs, goals, and programs they need to be focused on. For example the operations department will need to understand and focus on the KPIs and goals they impact, the sales and marketing department focused on the KPIs and goals they impact, etc.

In the scenario where an organization has one division but multiple facilities, the cascading is rather straightforward. The purpose, vision, and pillars should be cascaded directly down to each facility to ensure everyone is pointed in the same direction and focused on the right stuff. Since the KPIs measure performance under the pillars and progress towards the longer term vision, it is important that each facility, at a high level, is measuring the same stuff. For this reason, the KPIs should also be cascaded directly down to each facility so all facilities are using the same high-level measures to gauge overall performance.

Furthermore, since the goals are the catalyst for driving employee efforts and all employee efforts need to be pointed in the same direction to create alignment, the goals should also be cascaded down to each facility. Now, cascading the goals may be done in a few ways depending on the situation. At a high level, the cumulative performance of all facilities must achieve the overall KPI goals for the organization. For example, let's say a 5% improvement level has been set for one of the KPIs. Theoretically, if each facility realizes a 5% improvement then the overall 5% improvement will be realized for the organization. However, chances are that each of the facilities are performing at different levels so when cascading the goals there may be some modifications required. For example, take something like energy performance. If an organization had a goal to improve overall energy performance by 5% and one facility has already completed a number of initiatives to reduce energy, they may be given a smaller goal of say 2% or 3%. Whereas, another facility may have done nothing in this area and therefore has more opportunity for improvement and therefore are given a higher goal of say 7% or 8%. Nonetheless, the net impact across all facilities must realize the 5% goal for the organization.

Once the goals have been cascaded to each facility, it is then up to each facility to develop the programs for how their facility will achieve their portion of the goal. This also means that each facility will need to develop their own set of metrics and goals for their specific programs. The end result is that while each facility will have their own programs with specific metrics and goals to support these programs, everything that facility is working on feeds the higher level KPI goals and thus the strategy is aligned. It is recommended that each facility use the same process as presented in Chapter 7 to develop their specific programs, program metrics, and program goals. This will help ensure consistency in the approach and overall alignment of the strategic plan.

8.3 Building accountability

Accountability is vital to successful execution of the strategic plan. Many organizations struggle with this which negatively effects execution. Accountability is not about some draconian way of management or running a dictatorship; it is about setting clear expectations for performance and holding people responsible for achieving that performance.

The key to building accountability is having a *mechanism* for holding people accountable. This is where KPIs, Metrics and Goals come into play. Simply put, goals set the level of performance that is expected while KPIs and metrics measure the performance achieved – if desired performance is not achieved, the person who owns the KPI must answer. It is rather simple really, but many organizations fall down in this area due to lack of clear Goals and a lack of measurement. For example, setting a vague expectation or goal of "improving" or "reducing" does not clearly lay out expectations and makes it almost impossible to hold people accountable. Some people may think improve means 10%, others may interpret it as 1%. Having clear goals and ways of measuring progress towards those goals is critical to building accountability.

Fortunately, by having followed this process, the foundation for building accountability has already been laid. Clear KPIs for measuring progress and clear goals or commitment levels have been set. Programs have been put in place to support the KPIs and have their own set of metrics and goals. This allows accountability to be built at all levels of the organization, from the C-suite through to the front line.

The C-suite tends to be held accountable for the higher level KPI commitments whereas middle managers tend to be held accountable for goals related to the programs, with the front-line folks being held accountable for completing the actions laid out in the detailed action plan for each program.

Now, while KPIs, metrics and goals are the foundation for building accountability, it doesn't stop there. A system needs to be put in place that rewards or recognizes people for achieving the expected level of performance and penalizes people who do not achieve the expected level of performance. If there is no reward or penalty then there is no accountability. If a person does not achieve a goal, perform a task, or fails to follow protocol that leads to not achieving goals, etc., and nothing happens, it sends a very clear message that over time builds a very clear culture. The message is that nothing will happen if performance is not achieved, which breeds a culture of who cares, nothing will happen, so let's just carry on as usual. It is up to the individual organization to determine the reward, recognition, or penalty, but it must be there.

Now, a great system for rewarding and penalizing performance that will build accountability already exists in most organizations. Performance management systems provide a sound framework for managing performance and building accountability. The KPI commitment levels for each pillar can be tied into the performance management system for senior executives or managers. Goals related to the programs designed to achieve the higher level goals can be tied into the performance management system for middle managers. Completion of the smaller actions required to achieve the program goals can be tied into the performance management system for front-line employees.

In order for this to work properly, it is essential that the performance requirements not only be put into the performance management systems at each level across the organization but they are clearly communicated to each employee. If the organization has clearly communicated the strategic plan, employees will already have a high-level understanding of the goals, why there are important, how they were set, and so on. At this point, managers need to ensure that all members of their team clearly understand what specific goals and performance levels are required by them personally.

Additionally, it is critical that organizations have clear policies and procedures for dealing with performance. In other words, the organization must clearly define the reward(s) or recognition for good performance and the repercussions for underperforming. Communicating this in addition to the actual performance levels will go a long way in building a culture of accountability.

Once performance expectations are built into the performance management systems, regular performance reviews must be completed. An annual review is simply not going to cut it – the horse is already out of the gate by this point. At minimum, performance reviews should be completed twice per year. However, it is recommended that more frequent performance reviews be completed such as quarterly reviews. These more frequent reviews can be pared down to expedite the process.

The reviews should be a top priority and involve a one-on-one dialogue between the leader or manager and the team member who is being reviewed. The goal of these reviews should go beyond simply assessing whether goals or performance levels have been achieved or not and get into deeper discussions around the following:

- What worked well and what didn't?
- What challenges were faced and how were they overcome?
- Are the goals realistic?
- Are there any potential issues that could jeopardize achieving the goals going forward?
- Are there sufficient resources in place to support achieving the goals?
- Is any training, further support, or mentoring required?

Clearly this is not an exhaustive list of items to discuss in the review but rather some suggestions to get it started. The end goal of this review should not only be reviewing performance but also understanding the reasons and root causes for performance. This will enable course corrections to be made in order to achieve the desired level of performance as set out under the strategic plan. Through consistency and repetition of these reviews, a culture of accountability will slowly emerge and with this, the strategy will come to life and amazing things will happen as the vision becomes reality.

8.4 Managing strategy execution

It is not enough to communicate the strategy and build expectations into performance management systems. Organizations that are successful with execution, actively manage it. Borrowing from approaches used by leading organizations, here are some proven tactics for actively managing strategy execution:

1 **Build a reporting mechanism**: It is suggested that organizations build a means for reporting out performance under the strategic plan. Visual reporting via the use of visual boards works the best and there are different "levels" of visual boards that can be developed:

- High-level strategic board: A high-level strategic visual board can be set up for the executives and/or senior management team that is responsible for the high-level execution of the strategy. This board will include the high-level dashboard that was developed in Chapters 6 and 7. It should also include the purpose, vision, and pillars.

- Programs board: A visual board showing each of the different programs and corresponding goals and metrics should be set up in the area(s) where the program is taking place. For example, if there is a program related to customer service, the details should be placed on a visual board in the customer service area with the metrics and goals for that specific program included on the board. If the program relates to manufacturing, it should be set up in the manufacturing area, and so on.

2 **Performance reviews**: Once the reporting mechanism such as a visual board has been set up, regular report out and reviews must take place. The purpose of these reviews is simply to understand the current state of performance relative to desired performance. These meetings are simply a review, so they should be quick and to the point, lasting no longer than 10–15 minutes. The person responsible for the particular performance area should do the report out. This means that there could be multiple people reporting out at a meeting. By doing this, it creates another level of ownership and accountability for the goal, program, or action items.

During these review meetings, the overall leader or manager should be taking notes about the follow-up discussions that need to take place. For example, if it is a report out on the high-level strategic goals, the President or CEO should be taking notes. If it is about goals related to the various programs in place, it should be the manager responsible for the overall programs or the overall department taking the notes. The follow-up discussions should center around any particular goals or actions that are not meeting expectations along with things that are going well and exceeding expectations. The purpose of this follow up is to make real-time course corrections to actively manage performance and help ensure that all goals are achieved. By using this approach, it keeps the report out meetings quick and concise so people's time is not wasted while also allowing for further discussion and corrections to be made.

It is suggested these meetings follow a rhythm or cadence in that they are on the same day, at the same time, in the same room, and never missed. This is critical as it sends a clear message that the meetings are important and, furthermore, the strategic plan is important.

3 **Supporting the action plans**: With a system set up to report out
   and review performance, there must be a system in place to drive
   this performance. This is where the action plans that were developed in the previous chapter come into play. It is by completing
   each of the actions in the individual action plans that stuff actually
   gets done and results are realized. However, in order to move these
   plans forward and get the actions completed, they must be actively
   supported or managed. There is no secret to doing this effectively,
   it is fairly straightforward. Managers must meet with each of their
   team members on a regular basis to review each action in the plan,
   discuss any issues, work out solutions, remove roadblocks, and
   provide resources to ensure their action plan(s) is moving forward
   as scheduled. This simple technique is very effective as it engages
   the employee and addresses issues and problems immediately so
   there are no surprises at report out time. There is one trick to making this work – it actually has to be done, consistently, all the time.

8.4.1 Building the right team

Arguably one of the most common attributes of organizations who
are successful with execution is having the right people in the right
spot on the bus. This is a major differentiator between the great
organizations and the mediocre ones. It is kind of like a sports team.
Put the right combination of people together in the right positions
and magic happens. Have the wrong people or even the right people
in the wrong spots and mediocre performance ensues. For whatever
reason, too many organizations settle for mediocre performance and
hold on to employees who do not perform or hire people who are not
the right fit for the team. This mediocre performance acts like a can-
cer, spreading throughout the organization creating a culture where
mediocracy is not only tolerated, it becomes the norm.

Today, successful organizations are slow to hire and quick to fire.
These organizations do not tolerate mediocrity and at the same time
foster high performers. If a team member is consistently not performing even after mentoring or they do not fit the culture, they are
quickly removed and eventually replaced with the right person. This
sends a clear message to the rest of the organization that mediocre
performance will not be tolerated and over time, helps foster a high
performance culture. This is not an HR book and thus does not cover

hiring the right people and building a high-performance team in great depth, but rather suggests that this is an area critical to successful execution and it should receive due attention.

8.4.2 Being disciplined

Another very common trait that differentiates the great organizations from the mediocre ones is discipline. Successful organizations set a plan and stick to it – day in and day out, without exception. Like martial artists or Olympic athletes, they are focused on the end goal, they train every day, and practice the same moves over and over again until they are built into the muscle memory. They do not give up, they do not take a day off, and they do not lose focus. These same traits are present in successful organizations. The successful organizations do not stop communicating the plan, they do not miss a review meeting – ever, they do not stop managing the action plans – ever, they do not tolerate mediocrity – ever, they hold people accountable – always, and they never lose focus on the end game – in this case the vision and ultimately fulfilling the purpose.

There is no real secret to being a disciplined organization, it really comes down to the leaders and individuals in the organization. Like Olympic athletes, they must have the drive and determination to succeed, self-control and will power to make the right decisions, and do what is required even when they do not feel like it or are having a bad day. To make another analogy, it is kind of like the Karate Kid and Mr Myagi. A routine must be set up and relentlessly practiced day in and day out until it becomes second nature. Over time, the skills and abilities to execute become second nature but it requires constant discipline to get there and stay there. Without the leadership and coaching of the leader (Mr Myagi), the Karate Kid would never have been successful – on other hand if he was not determined and driven, it wouldn't matter what Mr Myagi did. This same thinking applies to organizations; the leaders must drive the discipline, set the routines, and provide coaching and mentorship for employees. However, if employees do not have the drive to be successful, it won't matter what the leadership tries, things will never reach the level of success that could be attained through committed employees – this goes back to having the right people in the right spot on the bus.

Through this rigorous approach, it turns what seems impossible or far-fetched at first (like catching a fly with chopsticks) into something that is plausible and eventually realistic. Furthermore, it builds a culture where discipline is just how the organization operates day in and day out. Organizations like General Electric and Toyota are famous for high levels of discipline and staying the course. False starts or flavors of the month are rarely heard of at these types of organizations. It does not happen overnight though; organizations that are disciplined and successful with execution have built the skills and "muscles" required over many years, even decades, of training, exercise, and preparation.

8.4.3 Locking in the gains

Part of successful strategy execution includes locking in the gains that are realized from deploying the strategy. Embedding improvements to sustain the gains can be done through proper governance. Governance refers to the development, management, and enforcement of an organization's policies, procedures, and systems.

Let's use a simple example where part of a strategy includes a focus on improving energy performance. When it comes to *policies*, they can be used to lock in the improved energy-related behaviors. For example, say it is determined that procurement behaviors must change to purchase more energy efficient equipment. Instead of assuming that this will always be part of the procurement process, lock this improved behavior in by adapting the procurement policy to make it a requirement. As long as the policy is enforced, this improved behavior is now locked in.

Continuing with the energy performance example, *procedures* can be used to lock in the improved operational practices. In this specific example of energy conservation, say one improvement implemented is to audit the air compressor lines for leaks and repair any leaks identified. This improved practice must be locked in somehow or else leaks get fixed once, short-term gains are realized but then going forward, this practice doesn't continue, new leaks occur, and the gains are lost. Developing or modifying maintenance procedures so that the air-line audit is now completed as part of regular maintenance will lock this improvement in.

Finally, *systems* can be used to lock in all types of improvements including improved behaviors, practices, and processes. Like procedures, systems come in many forms. They could be a formalized management system such as ISO 14001 or ISO 9001. A system could also be a performance management system as discussed earlier or even software and technology systems.

Management systems such as those offered by ISO are a great way to lock in improvements because often times they are a collection of smaller policies and procedures that can be modified to include the improved ways of behaving or operating. Performance management systems were discussed in more detail earlier as a means to build accountability and drive performance but they can also be used to lock in improvements. Since part of these performance systems often lay out job responsibilities or performance expectations, the improved behaviors can be embedded into a job description or improved performance could be set as the minimum performance requirement going forward, i.e. a quality rate of 98% must be maintained at all times.

Software and technology systems are also a great way to lock in improvements. Continuing with the simple example of energy performance, maybe room temperatures were dialed back as a means to conserve energy. Instead of relying on manual set back of temperatures on a daily basis, the new temperature settings can be programed into the thermostat or a building management system so they become automatic and can be sustained.

Obviously, this is just touching the surface in terms of how governance is used to lock in improvements, but the concept should be clear – improvements realized by deploying the strategy must be locked in to continually move organizational performance forward. Failing to lock the improvements in will result in a one step forward one step back situation, leaving the organization spinning its wheels and going nowhere.

8.5 Leveraging technology

In today's day and age, one cannot talk about successful execution without touching on technology. A word of caution here though, it is always best to learn how to multiply before using a calculator. In other words, learn how to properly execute first and then seek out the technologies available to support and expedite the execution. Having said that, there are many great technologies available that will help with execution and new ones emerge every day. A smart organization should always be on top of the existing, new, and emerging technologies that are out on the market whether they relate to strategy execution or any other aspect of the business. Showing up to a gun battle with a knife is not a gamble organizations want to take in today's fiercely competitive marketplace.

So clearly, execution is more art than it is science and there are many pieces to the execution puzzle. However, there are very clear traits and characteristics shared by organizations that are successful with execution and many of them are listed in this chapter. At a high level, the common thread amongst successful organizations is that they all have the basics down pat and build from there. While the things listed here are not the be all and end all, they work and are proven. Learning from the best practices of leading organizations and mimicking their behaviors will surely enhance execution for any organization.

Progress indicator

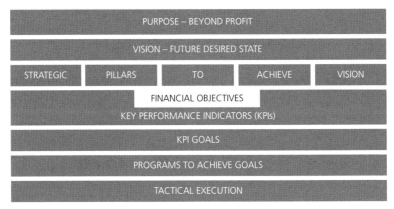

Fig. 8.1 The final step in the strategic planning process is complete – an execution plan is in place to successfully execute the strategy.

Final Thoughts – Making the Leap

Change before you have to.
Jack Welch

The system presented in this book aims to help organizations of all backgrounds, sizes, and types optimize performance through a more effective and agile approach to strategic planning – hopefully this has been accomplished. It should be obvious by now that the strategic planning system laid out in this book is not about developing a sustainability strategy but developing a strategy that is sustainable. There is a big difference between the two. A strategy that is sustainable is much broader than a sustainability strategy; it identifies all the factors critical to success not just ones related to sustainability, it focuses the efforts of employees to create the alignment needed to fire on all cylinders, it has the flexibility to quickly adapt and respond to market changes, and it provides the means for measuring performance and building accountability, all while maximizing positive impacts on the environment and society.

Clearly sustainability is not some passing trend or fad – it is here to stay. It is a critical component to competing and winning in today's

Purposely Profitable: Embedding Sustainability into the DNA of Food Processing and Other Businesses, First Edition. Brett Wills.
© 2016 John Wiley & Sons, Ltd. Published 2016 by John Wiley & Sons, Ltd.

economy. It is not the be-all and end-all for success though; many other ingredients go into a successful strategy. Nevertheless, there is no question that things have changed over the past decade or so and a successful strategy today, includes sustainability. Organizations across the globe have proven this over and over again through superior performance. Organizations like Unilever, Nestlé, Wal-Mart, Google, Marks and Spencer, Kellogs, GE, Virgin, Royal Bank of Canada, Duke University, Harvard, UPS, Walt Disney, and Apple have not built sustainability into their strategy because it makes them feel warm and fuzzy inside. They do it because they get it – sustainability is a better way to make a bigger profit. Sure, they may have their struggles like every other organization but there is no question that a focus on sustainability boosts their performance.

Organizations that resist this change will eventually be forced to change, whether through competitive necessity or other external forces – moreover, this will happen sooner rather than later. Unfortunately, organizations that have yet to embrace this change are already behind the 8 ball but following this system will definitely help these organizations quickly catch up. For organizations who have already embraced sustainability, this system will be the impetus for accelerating performance to reach new heights.

Let's be clear about one thing here. This system is not the brainchild of one individual. It comes from the minds of multiple individuals and the practices of many successful organizations. Distilling all this information down and organizing it has produced a simple but powerful and proven system to maximize performance. It is not an invention but an innovation which builds on successful and proven approaches that have stood the test of time while incorporating new thinking and approaches that are paving the way for the future.

It is not a quick fix or some miracle system though. It requires effort to develop and discipline to execute. Organizations that are serious about improving performance and committed to making it happen will find this system of great value. It provides the framework and proven approach needed to support a drive to do better.

Few people can read a book and immediately digest everything that is presented. If there is anything to be taken from this book, remember the eight points that follow. They will be beneficial for your organization like they are for many others.

1 Possess a reason beyond profit maximization for opening the doors every day. A genuine purpose is the stimulus that separates great organizations from the mediocre ones.
2 Have a clear vision of the future – it provides the direction and the point of alignment for everything else. Ensure everyone shares this vison and understands how they contribute to it.
3 Understand the organization's differentiators and elements critical to success. Use these to keep the organization focused and aligned.
4 Be able to measure performance. It is the catalyst for driving performance.
5 Give people something to shoot for. Goals are powerful motivators both in business and in life.
6 Embrace the spirit of sustainability. Doing so will open up a whole new realm of opportunities.
7 Build a team of all star performers. Today's competitive marketplace has no room for mediocrity.
8 Blueprints do not build houses. Designing the strategy is important but things only happen when the shovel hits the ground.

Of course, following the process and working through each step as laid out in the pages of this book will support all of the above. It will produce a strategy that will position the organization for success in a fiercely competitive global marketplace that is constantly changing and evolving.

Adapting the strategy

A strategy update will be required periodically and the update can be driven by a few factors. Responding to inevitable changes in the marketplace may require the strategy to be adjusted to address those changes. Scheduled updates will also be required to keep the strategy current. Fortunately, the model and overall system is designed to be agile and support rapid strategic changes. In order to revise the strategy, whether driven by market changes or simply a scheduled update, the following steps are suggested:
1 Purpose and vision: While these two foundational pieces of the strategic plan are long term aspirations that set overall direction and therefore should remain relatively static compared to the rest of the strategy, it is pertinent to review them regularly. The review

should focus on ensuring that the long-term aspirations are still relevant as well as fine-tuning the statements to provide further clarity that will foster alignment and culture development.

2 Business objectives: Ideally these objectives should be updated annually or when the time frame for achieving the objectives has expired. Of course, major disruptions or other circumstances may require more frequent updates or modifications to this set of indicators and goals. It is discouraged to simply modify objectives before the completion date just because the desired performance is not currently being achieved and there is fear that the objectives will not be met. Simply downgrading the objectives so they match what will likely be achieved defeats the whole purpose.

3 Pillars: Like the vision and purpose, the strategic pillars also represent longer term aspirations. However, since the pillars are a slightly lower level in that they represent the critical elements for how the organization is going to achieve the vision and fulfill the purpose, they are a little more subject to change. A major shift in consumer behavior or customer priorities, for example, may trigger a modification or complete change in one or more pillars. It is suggested to review the pillars annually along with the vision and purpose to ensure they are still relevant.

4 Key performance indicators: The KPIs are meant to measure performance and therefore should remain constant except in a few situations. Whenever a pillar is changed or modified, the KPIs should also be modified accordingly by repeating the KPI development process as laid out in Chapter 5. Furthermore, if it is determined that the KPIs are not the right ones, they should be modified or fine-tuned. Other than that, the KPIs should remain static in order to provide apple to apple comparisons year over year.

5 Goals: Goals are relatively dynamic and should be updated similarly to how the business objectives are updated – annually or whenever the time frame for the goals has expired. There should always be a current goal in place for each KPI.

6 Programs: The programs are also dynamic and because they the support the goals, they should be updated or at least reviewed whenever the goals are updated to ensure they will drive achievement of the goals. Additionally, programs should be updated if they are found to be ineffective in achieving the goals.

To summarize, the strategic plan should be updated annually. While the purpose, vision and pillars will be reviewed at this time, more often than not, they will remain constant unless there has been a major disruption to the organization or industry such as the change in consumer behaviors or customer priorities as mentioned earlier. Similarly with the KPIs, they should be reviewed annually to ensure they are the right ones to measure performance but unless something is awry, they will remain constant as well. The business objectives, the goals and the programs will be the main areas that change in the annual review. To do this, simply repeat the process for each that was undertaken to develop them in the original strategic plan. Of course, more frequent updates may be required if goals, programs, etc., expire before the annual review.

The modular design of this strategic planning system allows small changes to be made quickly without retooling the entire strategy. It also allows for more significant changes to be made in the form of pillar modifications. These pillar modifications will rapidly change the course of the organization to keep it headed towards the longer term vision and purpose. From a competitive perspective, strategic agility in and of itself is a competitive advantage. Some organizations can take months or more to make strategy modifications and adjust direction.

It should also be mentioned here that ongoing communication as discussed in Chapter 8 is key whenever any changes are made to the strategy. Folks must understand what changes have been made, why those changes were made, and how the changes will impact them.

What now?

An investment of time has been made in reading this book and now it is time to realize a return on that investment. The rub is that like after reading any book, attending a workshop, or training, it can be confusing or even daunting to figure out what to do with it all and what the next steps are. Here are some thoughts to help ease the transition from theory to practice.

If you are a senior leader with the capability to influence the strategic planning process, you have to determine if this model would be of value to the organization. If there is perceived value here, how will you adopt it? Will you adopt this process in its entirety or will you simply use it to enhance the current Strategic Planning process? If adopting in entirety, the next steps are pretty simple; assemble the strategy team and schedule a date to start the planning process, then work through each step in the process as detailed throughout each of the chapters.

For those using the system to enhance the current strategic planning process, thought must be given to what aspects of this system will enhance the current approach and how those will be blended into the current approach. What is missing in the current approach? Is it a particular piece of the Model such as the pillars or maybe the KPIs? If so, thought must be given to how this will affect the current pieces already in place. For example, if pillars are being added to the process but KPIs already exist, will those KPIs correctly measure performance of the pillars or will they need to be changed as well? Chances are, they will need to be changed.

If all the pieces already exist but maybe they are not aligned properly, then why is this? Maybe it is because the sequence in which the pieces are developed is out of order, so simply adopting the order presented under this system will correct alignment. Perhaps it is the execution piece that can use some tuning up, so adopting some of the techniques covered in that chapter will help. Regardless, once the problem and solution have been identified, be prepared to incorporate the modifications during the next round of strategic planning.

In the case where you do not currently have the capability to influence the strategic planning process, the information in this book can still be leveraged. One way to do this is simply adopt the system to develop a strategy for your own area, department, or team. The process is really the same except everything is toned down a bit. For example, the vision will be the future desired state of the area or the department or the team instead of the entire organization. The pillars will be areas critical to the department's success instead of the entire organization's success, so on and so forth.

Another way to leverage the information in this book would be to present it as an option for further consideration to those who do have influence with strategic planning. There are many different ways to go about this. Maybe it starts by simply sharing this book with them or preparing an executive summary. It could involve developing a presentation to make the case for adoption with a glimpse of what the strategy could look like by adopting the system.

Whatever is done, do something. To invest the time, even though not a huge amount of time, to read this book and do nothing, would be a wasted investment. Even if it is just thinking about the key takeaways or "aha" moments that resulted from reading the book, hopefully you take something away. In addition, do not give up after a few setbacks, stay the course. Use any bumps in the road as an opportunity to learn and improve rather than seeing them as a road-block and giving up.

Succeeding in today's global marketplace cannot be left to chance – it requires a sound strategy. One that incorporates sustainability, is driven by a higher purpose, and aligned to deliberately produce a profit. Organizations who use this model to build a strategy that is sustainable will optimize their ability to succeed and become purposely profitable.

Tools and Resources

There are numerous tools and resources available to help support and drive the strategic plan. From traditional management tools such as problem solving to standard continuous improvement or efficiency tools such as LEAN or Six Sigma to specific Sustainability related tools such as Biomimicry or Life Cycle Analysis, a multitude of tools and resources are available to support organizational performance.

The rub is that there are so many tools and resources available that it can become overwhelming. The trick is applying the right tool to the right problem or opportunity. Too many organizations arbitrarily or improperly apply tools and resources on the advice of a consultant or because many others have adopted it, and end up experiencing less than desired results. This approach is like a mechanic walking in to work in the morning, going over to the tool box, picking up any random tool, and going to work. Instead, a good mechanic will diagnose the issue then go to the tool box to get the right tool and then get to work.

When choosing which tool or resource to adopt, organizations must use this same simple philosophy of understanding the problem and then picking the right tool for the job in order to maximize performance. The trick is that in order to properly apply the right tool to the right problem, organizations must first have an understanding of the various tools and resources available and what each is designed to achieve.

Purposely Profitable: Embedding Sustainability into the DNA of Food Processing and Other Businesses, First Edition. Brett Wills.

This appendix provides an overview of the major tools and resources available to support the strategic plan. These tools and resources can be used to develop or drive the various programs or to support KPI and metric development or just provide general support in managing execution of the strategic plan. The point here is not to provide an in-depth review of each tool and resource but to introduce them and provide a basic understanding of what they are designed to do. From this information, organizations can better narrow down the tools and resources that could work best for their situation and explore them further. The tools and resources are sorted into different categories to provide a quick go to reference. While by no means an exhaustive list of the tools and resources available, it provides a good overview of what is available for further exploration.

Continuous improvement and operational efficiency

There are several tools and resources available that support continuous improvement and drive operational efficiency. These tools can be used to help identify and develop different programs for achieving the goals set under each of the Pillars. For example, if there is an efficiency goal under the Operations pillar, an organization may choose to focus on implementing a 5S program to help achieve this goal.

Lean: Lean is an overarching operational philosophy designed to identify and eliminate waste from the perspective of the customer while also focusing on adding value to internal and external customers. It is based on the identification and elimination of the 7 Lean wastes of: Transportation, Inventory, Movement, Waiting, Overproduction, Overprocessing, and Defects. Lean is widely adopted in some form or another by all types of organizations across the world to drive efficiency and effectiveness through the elimination of waste and the addition of value for the customer. This "tool" or philosophy can be used as a program to support the achievement of a number of pillar-related goals related to efficiency and overall continuous improvement.

There are also a number of tools, techniques, and resources that fall under the Lean umbrella which organizations can use to support an overall Lean journey. Let's explore the main ones.

Value Stream Mapping: A value stream is all the activities or steps (both value adding and non-value adding) required to bring a product or service from supplier to customer. A value stream map is a process mapping technique that identifies all of the lean wastes that are present within the value stream or process. This is typically the first step for organizations going down the lean path as it identifies what lean wastes are present in the process, where they are located, and to what extent they exist. This tool may be used to identify the source of inefficiencies that need to be addressed. From here organizations can adopt various other Lean tools listed later to build programs that will achieve efficiency related goals.

5S: This tool is a system for workplace organization. The system is called 5S because it is based on 5 Japanese words that translate to 5 English words starting with S: Sort, Set in Order, Shine, Standardize, and Sustain. In layman terms this is about having a place for everything and everything being in its place. Think about kitchen drawers. The cutlery drawer has a spot for everything – forks, spoons, knives. Because there is a spot for everything, all these utensils have a place and typically they are always put in their designated spot – it is super easy and quick to find a fork, spoon, etc. Now think of the drawer with all the spatulas, whisks, etc. There is not a spot for everything and thus not everything is in its place and therefore it takes much longer to find what one is looking for. The 5S tool is great for organizations that have a lot of wasted time and resources due to an disorganized workplace.

Kanban: A Japanese word for "signal", Kanban is an inventory control system designed to optimize inventory levels through just-in-time delivery of materials and supplies. In its basic form, this tool uses signal cards known as "Kanban Cards" that are physically placed on items to trigger or signal order of new materials. The cards list part number, supplier, reorder quantities, and other pertinent information required by the buyer to reorder. This tool works well for organizations that experience production delays or inefficiencies resulting from material delays or high inventory carrying costs. It is also a cost effective way to manage inventory in the absence of a software based inventory control system.

Gemba Walks: Denote the practice of going to see the actual process or work that is taking place. The purpose of the Gemba Walk is to better understand what is actually happening on the front line by observing the process and asking questions to develop solutions. The walks are designed to be done daily by leaders and are a great way to build culture through consistent positive reinforcement of practices valued by the organization. While not a solution per se to a particular problem or challenge, Gemba Walks support overall improvement and culture development.

Poka Yoke: Another Japanese term that means "error proofing." Ideally, Poka Yokes are mechanisms that ensure proper conditions exist before actually executing a process step to prevent defects from occurring in the first place. Where this is not possible, Poka Yokes provide a quality control function to eliminate defects as early as possible in the process. This tool would be well served by organizations which struggle with repetitive quality issues.

Andon: This tool is a visual feedback system for the production floor that indicates production status. Usually found in the form of a three-tiered light (red, yellow, green), this tool indicates when assistance is needed or there is a problem with a process and empowers employees to stop the production process when errors or problems are detected. Organizations which struggle with real-time communication on the floor would benefit from using this tool.

SMED: An acronym for "Single Minute Exchange of Dies", this tool employs a number of specific techniques to reduce set up time of processes and machinery to less than 10 minutes. A very effective tool for organizations which struggle with long setup and changeover times.

Kaizen: A general term meaning continual improvement, Kaizens are more often seen as a specific exercise of rapid improvement. A formalized problem solving and solution development process, Kaizens are set up to rapidly attack a specific challenge or problem using a proven step-by-step approach.

Heijunka: A philosophy for production scheduling that deliberately manufactures in smaller batches by mixing product variants within the same process. This tool works well for organizations that have long lead times and issues with overproduction.

3 Up Visual Board: This tool is designed to drive action and performance through visual boards or visual reporting that has three components. The first component or the first "Up" is reporting the trend vs the goal. While many organizations report out the trend which is great, many do not report it against the goal. Unless the trend is reported against the goal, it does not tell one if the trend is good or bad. The second Up, is the reason(s) for the gaps between the trend and the goal, this second up is also known as a type of Pareto Analysis. The third and final up is the action items to close the gaps between current and desired performance. These action items specifically address the reasons for gaps as stated in the Pareto Analysis. This third Up is the key to driving action through visual boards but requires the other two components to ensure the right actions are being worked on. This tool works well for any organization that struggles with effectively driving action through the use of visual boards.

Standard Work: A practice of capturing best practices for completing a certain task or process. An improved way to develop work instructions, this living document continually evolves as new and better ways of completing the task are developed. Developed with input directly from staff this tool is also a great way to capture tribal knowledge as a part of succession planning. For organizations that have quality issues from lack of consistency or challenges with effective work instructions this tool works well.

TPM: An acronym for "Total Productive Maintenance" this tool is an holistic approach to maintenance. The approach focuses on proactive and preventative maintenance to maximize the operational or "Up time" of equipment. Organizations that experience production delays due to equipment downtime or quality issues to due equipment malfunction, this proven approach to maintenance provides a great solution.

TWI: Another acronym that stands for "Training Within Industry", TWI is a dymanic hands-on approach for quickly and effectively teaching essential skills to those who direct the work of others. Developed by the US Military during WW2, this proven approach to training focuses on addressing the major skills gaps through Job Instructions (JI), Job Methods (JM) and Job Relations(JR). Organizations with issues resulting from inadequate training will benefit from adopting this approach.

Kata: A concept based on the repetitive sequence of movements practiced by martial artists, this tool uses disciplined routines to drive improvements. This structured approach to creating a culture of continuous improvement creates daily habits and routines that are practiced continuously until they become part of the "muscle memory" and are second nature to employees. Organizations which struggle with execution would be well served to adopt this tool.

Visual Factory: An approach that uses visual displays, indicators, and controls to improve communication of information, visually indicate optimal running conditions, temperatures, settings, etc. Organizations which struggle with quickly and effectively communicating information particularly around optimal settings or in situations where language is a challenge, will benefit from this tool.

Takt Time: This tool provides a method for calculating the ideal pace of production aligned with customer demand. For organizations that struggle with over or under production relative to customer demand would benefit from using this approach to determine what the ideal pace of production should be.

OEE: An acronym for "Overall Equipment Effectiveness" this framework is a method for measuring productivity loss associated with a given process. The measurement is based on 3 categories of losses: Availability, Performance, and Quality. In other words, 100% OEE would mean perfect production with no downtime, no defects, and ideal pace of production. Organizations which are experiencing or suspect that certain processes are not running optimally could utilize this approach to set a best practice benchmark for the process.

Six Sigma: This operational philosophy is a disciplined, data driven approach and methodology for eliminating defects. To achieve Six Sigma (six standard deviations between the mean and the specification limit), there cannot be more than 3.4 defects per 1 million opportunities. The Six Sigma process is driven by DMAIC – Define, Measure, Analyze, Improve, and Control. While relying heavily on Statistical process Control, Six Sigma like Lean, also utilizes a number of tools and approaches to drive improvement. Many of the tools are similar to the ones listed above and often organizations will combine the rigor of Six Sigma's statistical approach with the pragmatic approach of Lean to practice what is called "Lean Six Sigma."

Problem Solving Tools

The following tools and resources are specifically designed to solve problems. Some are more formalized and scientific while some are more abstract but all are designed specifically to solve problems or determine the root cause of a problem so a solution can be implemented. Again, not an exhaustive list with great depth of detail but rather an introduction to the most common and successful problem solving tools used by successful organizations.

The 5 WHYs: An iterative question asking technique that is used to determine the root cause of a problem. This tool suggests that by asking "Why" five times when analyzing a problem, the root cause can be determined. While five is suggested, sometimes the root cause can be identified in fewer tries and sometimes it requires more. Nonetheless, this is a simple but effective problem solving tool for everyday problems and issues that can be applied with little to no training on the tool itself.

Fishbone Analysis: Also known as "Cause and Effect Analysis", this problem solving tool is a diagram based technique to identify all the potential factors causing a particular event or effect. To help identify the causes they are typically broken down into six categories of: Equipment, Process, People, Materials, Environment, and Management. Once the cause of an effect is determined, other problem solving tools such as the 5 Whys can be used to determine what the root of the cause is.

A3: This structured problem solving approach was first employed by Toyota in Japan and is now widely used by Lean practitioners. Given its name by the "A3" size of paper it is printed on, this tool has seven areas that are completed to identify root cause and develop a plan to counteract the root cause. This tool is great for breaking down complex problems or issues and developing clear and concise countermeasures.

Kepner Tregoe: A problem solving approach based on eight questions developed by the consulting firm Kepner-Tregoe. This approach uses the process of elimination by asking seven high-level questions to pinpoint the cause of a problem. It starts with a clear problem description followed by these seven questions:

1. What is the problem? & 2. What is not the problem?
3. Where is the problem? & 4. Where is there not the problem?
5. When is the problem taking place? & 6. When is the problem not taking place?
7. To what extent is the problem?

There is a little more to this than just asking these questions but even by asking just these questions can help solve problems. The full scale approach to this tool would be useful for organizations with very complex and technical problems.

Sustainability tools and resources

The following tools and resources are focused specifically on driving Sustainability related performance, particularly in the areas of social and environmental performance. These tools can be used to help develop and/or support the programs required to achieve the Sustainability related goals under the different pillars. In some cases they may even be used to support non-Sustainability related goals such as improving product quality or efficiency by adopting a Biomimicry based approach.

Carbon Footprint: A carbon footprint, also known as a greenhouse gas inventory is a common tool used to help measure environmental performance. It is the amount of greenhouse gas emissions that are either a direct or indirect result of an organization, people, product, or event. The most common standard used to develop a carbon footprint is the GreenHouse Gas Protocol by the World Resources Institute.

ISO 14064: The internationally recognized system for managing the quality of a greenhouse gas inventory (aka Carbon Footprint). This management standard does not lay out the rules for calculating a carbon footprint but rather provides a framework for developing the footprint in conjunction with the Greenhouse Gas Protocol.

Carbon Disclosure Project: Also known as the "CDP" this program is the gold standard for reporting an organizations carbon footprint and carbon mitigation strategy. It is becoming increasingly common for large organizations to demand their suppliers report to the CDP. The reporting required under this program goes beyond simply stating the amount of carbon generated to ask questions around the level of commitment towards reducing carbon emissions, details on carbon reduction strategies, goals, and projects. Organizations which disclose through the CDP are awarded points based on answers to the questions. This is a great tool for organizations which want to be challenged to think deeper and more critically about carbon mitigation and management. The CDP has also come out with a new reporting framework specifically for water that follows the same approach as the Carbon Reporting.

Biomimicry: Biomimicry is the field of study that asks the question: How would nature do it? Popularized by Janine Benyus (2003) in her book *Biomimicry: Innovation Inspired by Nature* it uses nature as the inspiration to solve human problems. There are many famous examples of Biomimicry from swimsuits that mimic shark's skin to reduce drag, to the fronts of high-speed trains that mimic the beak of a kingfisher to eliminate the sonic boom when exiting a tunnel. This is a great tool for organizations that struggle with product, service, or process design.

Cradle-to-Cradle: Cradle to Cradle is a concept that was popularized by American architect Bill McDonough and German Chemist Michael Braungart (2009) in their book *Cradle-to-Cradle: Redesigning the Way We Make Things*. The concept is a design philosophy based on nature that eliminates waste and celebrates diversity. In layman terms it is about closed loop manufacturing where everything is designed to be upcycled at the end of its life or go back into the ground as a biological nutrient to grow more materials. Like biomimicry this is a great design tool but also a great tool for improving environmental performance of operations.

Sustainable Procurement: This enhanced form of procurement can be defined as the selection and acquisition of products and services that most effectively maximize positive environmental, social and economic benefits over the full life cycle. Going beyond the traditional drivers for procurement, sustainable procurement enhances the purchasing function by integrating Sustainability considerations into the purchasing practice. This is a great way to help embed Sustainability into daily decision making.

ISO 14001: Part of the International Organization for Standardizations' environmental series of management systems, ISO 14001 sets out the criteria for an effective environmental management system. While not prescriptive in stating requirements for performance, it lays out the framework for managing and continually improving environmental performance through a structured system. One of the most popular global approaches to managing environmental performance, organizations can become certified to this standard. This is also a great way to embed sustainability into the daily decision making.

Life Cycle Analysis (LCA): Similar but broader than a carbon footprint, this is a technique to assess the environmental impacts (beyond just carbon) of a product, process, or service throughout the entire life cycle. The life cycle includes all the steps from material extraction through to disposal or end of life and all steps in-between. This tool would be valuable for organizations which have exhausted all the low hanging fruit and are looking to further improve the environmental impact of a product, service, or process.

ISO 50001: Similar to ISO 14001, this standard is an internationally recognized framework for developing an effective management system focused solely on managing energy performance vs broader environmental performance. It is not prescriptive nor does it lay out fixed reduction targets, it is designed to provide a structured approach to continually improve energy performance and can be third party certified. While any size organization can adopt and benefit from this system, larger organizations tend to realize greater benefits. Often, organizations will simply integrate the framework of this system into an existing system such as ISO 14001 or ISO 9001. This is a great tool to embed energy conservation into daily decision making.

ISO 26000: This internationally recognized framework provides guidance on how organizations can operate in more socially responsible ways. While it does touch on the environment, it focuses more on the social side of Sustainability in areas such as Human Rights and Labor Practices. Set up as a guidance standard it is not auditable or certifiable by a third party. This tool will help organizations which are struggling with driving social Sustainability.

ISO 14040: This is ISO's standard for managing Life Cycle Assessments. It does not provide the technical detail of how to calculate life cycle impacts per se but rather provides a framework for ensuring the life cycle assessment is carried out and managed properly. It is auditable and certifiable by a third party.

Lean and Green: A systematic, step-by-step approach for improving environmental performance. Building on the traditional Lean philosophy of looking at an operation from the perspective of the customer, Lean and Green looks at an operation from the perspective of the environment. The foundation of the Process is based on Identifying, Measuring, Minimizing, and Eliminating the seven Green Wastes of: Energy, Water, Materials, Garbage, Transportation, Emissions, and Biodiversity. This is a great approach for organizations which are just getting started or which are further down the path and looking for a more formalized approach to improving environmental performance.

Green Value Stream Mapping: A take on the traditional value stream mapping process and used in tandem with the overall Lean and Green process. This process mapping technique identifies all the green wastes that are present throughout each process or step in an operation.

Zero Waste to Landfill: This is a concept that focuses on diverting waste from landfill. While there is no internationally recognized standard to define what determines "zero waste," at a minimum it is defined as 90% diversion. In other words, of the waste generated by an organization, a minimum of 90% of that waste must be diverted from landfill through recycling, re-use, composting, etc. A typical first step for organizations looking to improve environmental performance, organizations around the globe have adopted this philosophy and set waste diversion goals for their organization.

Waste Audit: This exercise is a formal, structured process for measuring the amount and types of waste generated by an organization. The audit results in the identification of opportunities to divert waste from landfill as well as opportunities for reducing the amount of waste generated in the first place. This is usually one of the first steps taken by organizations which are working towards zero waste to landfill.

Waste Brokering: A new type of business model that has recently emerged to support organizations in their waste diversion efforts. Waste Brokering is tool that adds value in a few different ways. First off, it streamlines waste management for organizations by managing all of an organization's waste haulers, recyclers, and compost providers. Secondly, they provide an outlet for recycling or re-using odd ball or hard to divert items that are not accepted by traditional waste haulers or recyclers. Finally, they are incentivized through a performance contracting model to increase diversion rates and realize cost savings by better diverting waste from landfill. This service is offered by organizations known as waste brokers.

LEED: An acronym that stands for "Leadership in Energy and Environmental Design", LEED is an internationally recognized Green Building Standard. Governed by country specific Green Building Councils this standard is a point based system that awards points based on exceeding existing building standards in a number of categories such as Indoor Air Quality, Energy, and Water. Buildings can be certified to different levels of LEED such as Bronze, Silver, Gold and Platinum. There are also other Green Building standards and programs available such as BOMA Best.

CSR Reporting: An acronym that stands for Corporate Social Responsibility reporting more commonly being referred to as Sustainability Reporting. This is the practice of publicly disclosing an organizations Sustainability related performance. Sustainability or CSR Reports typically present an organizations strategy for Sustainability, the aspects that are material to their organization, goals related to Sustainability, performance relative to the goals and shares stories, projects and overall efforts on their journey towards Sustainability.

Global Reporting Initiative: Commonly referred to as "GRI" this is the gold standard for Sustainability reporting across the globe. This protocol provides a standardized approach for reporting Sustainability performance. It is rather prescriptive in that it dictates what is to be reported, how to report it and even goes down into details such as the units of measure to use when reporting. This is a great tool for any organization to adopt that wants to lend credibility to disclosure of their Sustainability performance.

Integrated Reporting: This is an evolution of traditional corporate financial reporting that integrates reporting on Sustainability performance to provide a better overall picture of organizational performance. A protocol for integrated reporting is available through the International Integrated Reporting Council (IIRC). This is more sophisticated approach to reporting that is usually adopted by organizations with experience in both financial and Sustainability reporting but can be used by any organization.

Future Fit Business Benchmark: This tool defines a set of science-based performance criteria that describe a company that is fit for the future. In lamens terms, it defines the performance levels that are required to be a sustainable organization. Developed by 2 leading Sustainability gurus, this tool is practical and pragmatic and will help to set goals and comittment levels.

United Nations Global Compact: Known as UNGC, this strategic policy initiative is focused on aligning businesses to ten universally accepted principles in the areas of: Human Rights, Labor, Environment, and Anti-Corruption. It requires that organizations sign on and commit to adopting the ten principles and regularly report out on progress. While not all organizations may choose to sign on to this program it is a great tool for improving social performance.

Third Party Certification Programs: There are a large number of product and service Sustainability related certification programs available. These third party certification programs certify everything from energy efficiency to how the product was harvested or manufactured. These certifications can be helpful tool in a few different ways. First, for organizations looking to procure "greener" or more sustainable products or services, these certification programs help buyers choose the best solution. Secondly, manufacturers and service providers can have their own products and services certified under one or more of these programs as a means to drive increased revenue. It is important to note that not all these programs are equal, some are much more rigorous and thus much more accepted than others. It is imperative that organizations do their homework when choosing the program(s) to adopt.

Energy Audit: This exercise involves bringing in an engineer or certified technician to inspect, survey, and analyze energy usage of a process, system, or building for the purpose of identifying opportunities for energy conservation. This is typically an initial step for organizations looking to improve energy performance. Often this service is available through a local utility or local distribution company as well as third party engineering and consulting firms.

Water Audit: Similar to an energy audit, this exercise involves bringing in an engineer or certified technician to inspect, survey, and analyze water use within a process, system, or building to identify opportunities for water conservation. Like an energy audit, water audits are a good first step for organizations looking to improve water use. Often times this service is available through the local utility or water company as well as third party engineering and consulting firms.

EPI: An acronym for "Environmental Performance Indicator", this is a tool to measure environmental performance using a single number. Developed by a collaborative group of companies in Toronto, Canada, this tool combines energy, water, and waste performance using dollars as a common unit of measure. By translating energy usage, water usage and waste generation into dollars and applying a normalizer such as units of production, organizations can better measure and manage their environmental performance.

A final note ...

There are many other tools, resources, techniques, and so on available to organizations that will in some way support the strategic plan. The ones listed here are some of the most common, proven, and dynamic tools used by many successful organizations. This is a great place to start but organizations are encouraged to explore other tools if the desired solution is not listed here or not working in a particular situation.

Organizations must remember though that the key here is to ensure the right tool is chosen for the right job and it is applied in a disciplined manner – follow through and give it time to work, changes do not happen overnight. Organizations must stay the course and be disciplined – no false starts or flavors of the month.

Finally, there are many ways to gain further knowledge, information, and experience around the various tools and resources available. Traditionally organizations would turn to training, workshops, or a consultant and this works but there are other ways. There are multiple websites, blogs, social media outlets, etc., available online that have videos, templates, stories, instructions, etc. that will support further learning. Additionally, one of the best ways to really learn about the tools is to engage with an organization that has experience working with the tool. Whether it be through a formalized collaborative or informal best practice exchanges, this is a great way to learn the fine nuances of what works and what doesn't.

For more information the tools and resources listed here as well as many other supporting resources to support strategic planning please visit www.purposelyprofitable.com

References

Benyus, J. (2003) *Biomimicry: Innovation Inspired by Nature*, 2nd edn. HarperCollins.

Deloitte Development LLC (2014) *Culture of Purpose – Building Business Confidence; Driving Growth: 2014 Core Beliefs & Culture Survey*, 2, 10–11.

Doran, G.T (1981) There's a S.M.A.R.T. way to write management's goals and objectives. *Management Review* (AMA FORUM) **70**(11), 35–6.

Hawken, P. (1993) *The Ecology of Commerce: A Declaration of Sustainability*. New York: Harper Collins.

McCormack, M. H. (1986) *What They Don't Teach You at Harvard Business School: Notes from a Street-smart Executive*. Bantam Publishing.

WCED (World Commission on Environment and Development) (1987) *Our Common Future*. Oxford: Oxford University Press, p. 43.

www.thenaturalstep.org. The Natural Step. The Four System Conditions.

www.gamechangers500.com. Game Changers 500.

Index

*Purposely Profitable: Embedding Sustainability into the DNA of Food Processing
and Other Businesses*, First Edition. Brett Wills.
© 2016 John Wiley & Sons, Ltd. Published 2016 by John Wiley & Sons, Ltd.